Nadia Boulanger

NADIA REISENBERG: A Musician's Scrapbook

Robert Sherman
and
Alexander Sherman

International Piano Archives
at the
University of Maryland
College Park

Manufactured in the United States of America

Library of Congress Cataloging-in-Publication Data

Nadia Reisenberg: a musician's scrapbook

I. Reisenberg, Nadia. I. Reisenberg, Nadia.
II. Sherman, Robert 1931 – III. Sherman, Alexander.
ML417 .R4N2 1986 786.1'092'4[B] 86-72
ISBN 0-918512-05-0

CREDITS AND ACKNOWLEDGEMENTS

Our sincere thanks to the many photographers whose work has been reproduced in the following pages. In most instances, unfortunately, we found no name credits on the pictures, so we apologize in advance to those artists who must remain anonymous.

Among those we have been able to identify, however, and to whom, therefore, we can and do express our special appreciation, are James Abresch, Paul Affelder, George Faitzer, Larry Gordon, Chinho Kim, Fred Marcus, Zvi Oron Orushkes, B. Sladon, Thecla, and Luigi Pellettieri, whose beautiful portraits of Mother appear in the final chapters, as well as on the back jacket of the book.

Similarly, we are indebted to the many reporters whose reviews, interviews and feature articles about Mother form so important a part of this Scrapbook, and to the newspapers and magazines which commissioned them. By far the greatest number of these reprints are from the New York Times. Reviews and articles excerpted from the New York Times are copyright © 1924, 1927, 1933, 1937, 1941, 1944, 1945, 1950, 1951, 1952, and 1979. Reprinted by permission.

The first page of Samuel Barber's Opus 20, "Excursions," is copyright © 1945 by G. Schirmer, Inc. Reprinted by permission.

R.S. and A.S.

Produced for the International Piano Archives at the University of Maryland by North River Press, Inc., Croton-on-Hudson, New York.

Contents

Gary Graffman, Freda Berkowitz, and Nadia Reisenberg

Foreword

A Recollection of Nadia Reisenberg

I was never introduced to Nadia Reisenberg. Rather, I became, as a very young child, gradually conscious of her presence. It was a cheerful, smiling presence, joking in mellifluous Russian at her sisters', where I occasionally accompanied my parents for tea; it was a serene and dignified presence on the stage of Carnegie Hall, where I was often taken to hear her play with the Philharmonic, or the more intimate Town Hall, where she treated us to recitals of meltingly lovely Schumann, lucid Haydn, and demonic Prokofiev; it became, a little later on, a frequent, almost regular, and—thank God!—always compassionate presence at the studio of her friend (and my teacher), the volatile Isabella Vengerova, where young pianists like myself did the playing and were afterward exposed to merciless criticism. The mere fact of Nadia's being there, seated quietly on the long couch near the piano, her sweetly understanding eyes telegraphing enthusiasm as well as sympathy to the victim, somehow cushioned even the harshest verbal blows.

But kindliness is not necessarily next to flabbiness. Nadia, who over the years infected whole regiments of young pianists with her contagious vitality, her joy of discovery, her sheer love of music, was, as a teacher, a strict taskmistress in her own fashion. She expected hard work, unwavering discipline, and seriousness of purpose from her pupils. She set the highest standards. She just happened to go about her business in a gentle way.

Nadia's manner of dealing with young artists became familiar to me in later years when we served together on the jury of the Leventritt Competition. Sometimes there were as many as seventy young hopefuls in the preliminary auditions, most of whom would have to be eliminated before the next round. Since no point system was employed in the judging, lengthy, probing, and sometimes heated discussions took place instead. Nadia's analyses were never cruel, but always to the point: "Mr. X? He has good fingers—but to what use does he put them? That strange tempo in the slow movement of the Beethoven...and his shrill tone in the Chopin..." and so on, her lilting Russian accent somewhat modifying the severity of her views. Finally, a joint decision would be reached. "But Mr. X?" Nadia would ask. "He has been eliminated? How is this?" Ten pairs of bloodshot judges' eyes focused on Nadia. "But you yourself said..." Nadia: "The Beethoven. Yes, yes, I know. Of course, it was simply *drrreadful*!" Her hands flew to her face, protecting it from contamination. A sad smile followed, and then, pleadingly, "But perhaps he has not had the benefit of much experience—nerves? He must have *some* talent—after all, that section of the Liszt was quite remarkable, don't you agree? Such intelligent phrasing! Surely we must give him another chance to show what he can *really* do!"

Somehow, Nadia always managed to find a way to kindle sparks of enthusiasm and sympathy. And, ever since I can remember, these were among the sparks that illuminated her so radiantly as well.

Gary Graffman

Editors' Preface

Preparing this book of memories has been a voyage of fascination and frustration in almost equal measure. How wonderful it has been to sift through these extraordinary mementos of musical—and family—history, how many childhood recollections they sparked, and how proud they have made us (all over again) of Mother's achievements. On the other hand, why are there so many gaps? How is it, for instance, that in all the hours of taped interviews, we never thought to question Mother about her experiences with Alexander Lambert, her first teacher in America? Why can't we find such a major document as the Tarbut Medal she was awarded by the America-Israel Cultural Foundation in 1974? Who are some of the familiar faces (but, alas, forgotten names) in the older photographs?

What we have here, then, is not a proper biography, nor a complete chronology, nor even a fully researched documentary on the life, art, and times of Nadia Reisenberg. Those, we hope and trust, will be undertaken in due time by others. Rather, this is a scrapbook, affectionately recalling some of the high points, both professional and personal, of a warm, vibrant life in music.

About the text: we are lucky enough to have many reminiscences in Mother's own words, including letters, printed articles, and conversations taped at home, off the air, at master classes in Jerusalem and elsewhere. Unless otherwise indicated, these are the sources of her quoted statements. To fill in some of the missing links in her story, especially of the early years (and to provide translations of the Russian documents), we turned most often to Mother's closest relations and dearest friends: her sisters Anna Sherman and Clara Rockmore. To us, though, they are simply Newta and Clara, and it is thus that their contributions are credited.

We are deeply grateful to Mother's many friends, colleagues, and students (most especially Andrei Sedych, the noted writer and publisher of the Russian-language daily newspaper *Novoye Russkoye Slovo*) for helping us obtain some of the materials you will find here. Our special thanks go as well to William Cowan and Laurence Gadd of North River Press for their technical advice and moral support; to Steve Sherman who skillfully photographed some of the documents and reprocessed others; and to Neil Ratliff, Head of the Music Library at the University of Maryland, and his associate, Bruce Wilson, for their unstinting and enthusiastic support of the entire project.

We are, incidentally, proud and pleased to say that the International Piano Archives at the University of Maryland now contains a Nadia Reisenberg Archive, where other historic photographs, letters, and papers, along with tapes of as many of Mother's performances as we can locate, are being filed, indexed, and preserved. If you have other materials which you feel might be of value to the collection, or simply would like to learn more about it, we hope you will feel free to contact the Music Library at the University of Maryland, College Park, Maryland, 20742.

And so, let us dedicate this book to the memory of Nadia Reisenberg, even as Artur Rubinstein inscribed a copy of his autobiography for Mother: "to a fine lady and a wonderful pianist, with admiration." And, we might add, with abiding love.

Alex and Bob Sherman
1985

Nadia Boulanger

Family Album

PATERNAL GRANDPARENTS

Eve and Ossip (Joseph) Reisenberg

MATERNAL GRANDPARENTS

Chaim Grad

Sara Grad

PARENTS

Aaron Reisenberg

Rachel (Grad) Reisenberg

Ааронъ Іосифовичъ
Рейзенбергъ

Главный бухгалтеръ
Акц. О-ва „Лекторъ“.
Акц. О-ва „В. Гильванъ“.
Акц. О-ва „Біохромъ-Біофильмъ“.
Т-ва н/п Гимназ. М. Н. Стоюниной.
Т-ва Ф-ки І. И. Серебренникова.

радъ, В. О. Большой пр. 56. Тел. 548-79

AARON OSSIPOVICH REISENBERG
Certified Accountant
Lektor Co., Inc.
V. Rivlin Co., Inc.
Biochrom-Biofilm & Co. Inc.
Gymnasium (School) Stoyunninoy
Factory of I.E. Serebrianikov

Petrograd V.O. (Vassily Ostrov) Bolshoi Prospect (Ave.) 56 tel. 548-79

Rachel and Aaron

Uncle Paul Grad

I didn't begin playing until the age of six, for the excellent reason that we didn't have a piano. But then one arrived, as a present from a favorite uncle, and it soon became clear that I would be at the keyboard for the rest of my life.

NR, 1974

Newta and Nadia

I was brought up on Mother's milk and Nadia's practising. I never knew life without her piano. I would lie in bed and hear Nadia going through every Chopin Etude, one by one, and I knew them all by heart. The whole piano repertoire is very familiar to me to this day because she was always practising and I was always listening.

Clara, 1984

Clara, Newta, and Nadia

In 1911, I toured Russia and in December visited Vilna. When I mentioned to my uncle in Bialystok my intention, he asked me to bring his regards to the family Reisenberg there. This opportunity pleased me greatly, and upon my arrival the first place I went to was the house of the Reisenbergs, not dreaming what wonderful surprises were awaiting me there.

I rang the bell and a charming little girl, about seven years old,with a pleasant, friendly voice, opened the door. Seeing a stranger, she did not get confused, but asked me whom I wanted to see, and when I had identified myself, she very politely asked me to come in, to take off my hat and coat and put away my valise, and like a well-experienced hostess she led me into the drawing room. She invited me to sit down, and then she sat herself down, leading a conversation not a bit worse than any good hostess would do.

After a few minutes, she excused herself, walked out of the room, only to return a few minutes later with a tray loaded with tea, cakes, and sweets, which she served for both of us. She expressed her regret that her parents were out, but assured me that they would return soon. I watched the extraordinary child, who made me forget that only a child was in front of me, and I began to talk to her as to a grown-up person, and she very intelligently answered all my questions.

When we finished our tea, she asked me if I would like to hear some music. I expected to hear a gramophone record of some kind, but my young charming lady got up, walked over to the piano, opened it, and began to play. To play not as a child, but as a virtuoso!

I was so enchanted by her behavior and playing that I felt as if I had wandered into a beautiful dream, watching an act of rare beauty. I could have sat there for hours listening to her play, but all of sudden she stopped, and I woke up as from a trance. I looked around and there were her parents with two other girls, Anuta, age six, and Clara, about two years old, and my lovely hostess was introducing me to her family. Meeting her dear parents, it became clear to me from whom Nadia had inherited that gift of charm, tact, warmth, and kindness.

A reminiscence by Zvi Orron (Orushkes), Jerusalem, 1962

Newta, Aaron, Clara, Rachel, Nadia, and Grigory Reisenberg (Aaron's brother)

Even as a young girl, she was an amazing personality. She was never downcast, she was always an optimist, the strong one in our family. We looked up to her, she was the leader.

Clara, 1984

When I was a child, I had to read through all three movements of a piece even if I only had to prepare one for the lesson. I'm much too curious, so even before I do any thinking about a score, I want to read it from beginning to end and know what it's all about. Later I put everything together.

NR, 1981

Early Years

My parents moved to St. Petersburg just so I could study at the Conservatory there. I was then a funny-looking, rather unattractive little girl with braids and a great big bow.

NR, interview with Andrei Sedych, 1974

М. Г.

Настоящимъ имѣемъ удовольствіе сообщить Вамъ, наши дочери: 5-ти лѣтняя Клара 2-го с. м. отлично выдержала вступительные экзамены и принята, какъ исключеніе, въ Петр. Императ. Консерваторію по классу скрипки и 11-ти лѣтняя Надя съ отличіемъ перешла 12-го с. м. на высшій курсъ къ профессору Николаеву по классу рояли.

Съ совершеннымъ почтеніемъ

А. О. Рейзенбергъ съ супругой.

Петроградъ, 15 Сентября 1915 г. Вас, Остр. Больш. пр., 56. Телеф. 548-79.

Herewith we have the pleasure of announcing to you news of our daughters: 5-year old Clara, on the 2nd of this month, passed remarkably well the entrance examination and, as an exception, was accepted by the Petersburg Imperial Conservatory to study violin; and our 11-year old Nadia, on the 12th of this month, entered the higher course, with honors, for studies in the piano class of Professor Nikolaiev. With great respect,

A.O. Reisenberg, with his wife.
Petrograd, 15 September, 1915

When I was called to audition in the study of the director of the Conservatory, Alexander Glazunov, I also found there Professor Isabella Vengerova and the Inspector of the Conservatory. I played for them, and Glazunov leaned over to Isabella and said "Well, what do you say, shall we give her the 'plus' [5+, the highest grade]? She doesn't need any other exams, she had one right now."

NR, interview with Andrei Sedych, 1974

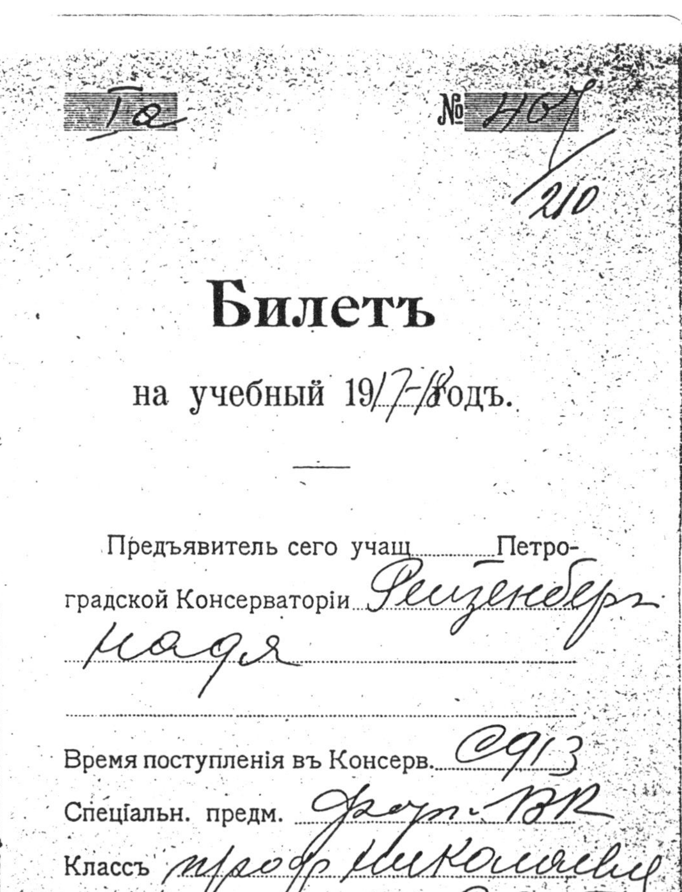

№ 407 / 210

Билетъ

на учебный 1917-18 годъ.

Предъявитель сего учащ..... Петроградской Консерваторіи Рейзенберг Надя

Время поступленія въ Консерв. 1913

Спеціальн. предм. Фортеп.

Классъ проф. Николаева

Время поступленія въ классъ 1915

TICKET [of Admission to Conservatory] for 1917-18 year

Student of Petrograd Conservatory
Reisenberg, Nadia
Time of first entry to Conservatory *1913*
Specialty, *Fortepiano*
Class, *Prof. Nikolaiev*
Time entered class *1915*

To my dear young friend, the splendid pianist, Nadia Reisenberg from

Isabelle Vengerova

Утро учащихся.

Участвовавшіе въ музыкальномъ утрѣ 6-го декабря представители фортепіаннаго класса профессора консерваторіи Л. Николаева проявили отличную подготовку къ концертной дѣятельности исполненіемъ программы, въ которую, большею частью, вошли труднѣйшія произведенія Листа, Глазунова, Шопена, Баха-Бузони, Баха-Листа. Среди шести выступавшихъ піанистовъ находилась и малолѣтняя ученица, игравшая съ сопровожденіемъ второго рояля (г. Николаевъ) «Каприччіо» Мендельсона—пьесу, конечно, не той трудности, какою отличаются остальные номера, но тѣмъ не менѣе, достаточно отвѣтственную для столь юной піанистки. Ученица Розембергъ выказала въ этой пьесѣ весьма развитую бѣглость и большое бріо.

Программа утра открылась прелестными и почти совершенно забытыми варіаціями Шопена на тему изъ оперы «Людовикъ» въ исполненіи ученицы Акоповой, сумѣвшей выдвинуть изящество шопеновскаго стиля съ пикантной ритмикой чудеснаго «скерцо» включительно. Далѣе послѣдовалъ истинный экзаменъ по части техники, оказавшейся у соперничавшихъ между собою исполнителей на томъ высокомъ уровнѣ, который требуется для виртуознаго исполненія испанской рапсодіи Листа (уч. Влашекъ), «Шаконны» Баха—Бузони (ученицы Гольдфельдъ), первой части концерта Глазунова (уч. Мацулевичъ со вторымъ роялемъ въ исполненіи Л. Николаева), «Тарантеллы» Листа (ученикъ Софроницкій) и листовскаго «Мефисто-вальсъ» (учен. Шварцъ). Юные піанисты увлекались порою въ сторону излишней силы удара и преувеличенной скорости темпа: вліяніе темперамента объясняется въ данномъ случаѣ, быть можетъ, повышенной нервностью участвующихъ въ публичномъ состязаніи.

Л. Николаеву были поднесены цвѣточные трофеи при единодушныхъ вызовахъ.

MORNING OF STUDENTS

The participants in the musical morning of December 6th, representatives of the piano class of Prof. Nikolaiev, showed a remarkable readiness for concert careers. They presented a program which contained the most difficult pieces of Liszt, Glazunov, Chopin, Bach-Busoni, Bach-Liszt.

Among the six performing pianists was a young student who played Mendelssohn's "Capriccio" accompanied at the second piano (by Prof. Nikolaiev), a work not as difficult, perhaps, as those performed by others, but nonetheless very respectable for such a young pianist. The student, Nadia Reisenberg, showed in this piece a well-developed technique and much vivacity....

L. Nikolaiev received an ovation, as well as several floral arrangements.

1916

My first impression of Nadia and her beautiful playing goes back to days long gone when she, as the youngest student of Prof. Nikolaiev, was entrusted to present a gorgeous bouquet of red roses to *his* teacher, Wassily Safonoff, whom Nikolaiev had invited to listen to his pupils playing at the Petrograd Conservatory. This was in 1917. I was there too, with my father. With loving thoughts of her...

Maria Safonoff, 1983

Almost everything I know about the physical side of piano-playing, the real Russian schooling with economy of movement and the art of legato, I owe to Nikolaiev's extremely detailed schooling. He was an incredible pedagogue, not so much in musicianship, which I developed afterwards, but in terms of giving me all the basics. He gave me that which has served me in all the years since. He made me aware that economy of motion is one of the most important things: If you watch me play you don't see my hands flying up in the air, going here and there and all over the place. I don't look busy. I'm busy with other things, with listening, with music...

NR, 1974, 1981

When I went to Nikolaiev, my thumb was very weak and fell in prominently. He eliminated this defect by making me use the thumb only on the edge of the tip, slightly in the upright position. This made it flexible and light, a necessity to any player, and above all to a Mozart player. He taught me how to let my thumb hang perfectly free so the fingers might move on unhindered.

In playing chords, Nikolaiev taught us to mold the hand into position as it attacks the chord—to avoid attacking with the fingers already set. Anything that is set is rigid, and if the chords are played with a fixed position in advance, every note will not sound. Feel each key as it is struck, and each note will sound true. He insisted on a simple, natural position of the hands and fingers, objecting to all showy, theatrical movements, such as lifting the hands high above the keyboard or making unnecessary movements after a sustained note has been struck. "You can influence the quality of sound by the way you strike the key," he told me, "but once it is struck, all wriggling around is pure mannerism."

A completely loose wrist, a proper application of arm weight, and the greatest economy of motion were the three Rs of my technical schooling.

NR, 1940

To Whom it May Concern
A student of my class, N. A. Reisenberg, I recommend as an extremely talented and, in spite of her young age, fully accomplished pianist and excellent musician.

Professor of Petersburg Conservatory
L. Nikolaiev 28 August 1918

Предъявительницу сего, ученицу моего класса Н. А. Рейзенбергъ, рекомендую какъ чрезвычайно талантливую и, несмотря на юный возрастъ, зрѣлую піанистку и отличнаго музыканта.

Профессоръ Петербургской Консерваторіи

Л. Николаевъ

28 августа 1918.

My beloved Nadusha,
I bless you on your artistic voyage.
From your old teacher and friend.

I find teaching exciting, gratifying and rewarding, and it's always been that way. I remember that even as a youngster in the Conservatory I was forever being asked to help other students with theory problems or four-part harmony exercises. "Come on Reisenberg, it takes you five minutes and I'll be working an hour at it." And I'd say "No, I'll show you how to do it, but I won't do it for you." Even then I realized that the point is not just to be able to accomplish something, but to know why you're doing it, and how it should be done.

NR, 1979

Our two uncles—Uncle Paul and Uncle Max—kept writing us constantly from Vilna to say that things were a little better there, that they had practically a whole cellar-full of potatoes, and at least the children should try to get there. When we left Petrograd, it was not with the idea of moving permanently; it was for the summer only, while there was no school and the Conservatory was also closed, just in order to be fed a little more. Father stayed behind—he was in the Red Army—and he used to write that he was preparing things, trying to find a little butter and so on, for when we returned. But before that happened the situation changed, and there was actual war going on in the streets of Vilna. Practically every other day a new regime took over, and there was fighting going on all around. Our uncles had a big drugstore, and soldiers used to come in and smash everything and scatter things around the floor. We would hide in the cellar where the potatoes were, and sometimes they'd find us and they'd pull us all out and search us. Father realized it was impossible for us to get back to Petrograd on our own in this situation, so he got permission to go to Vilna to bring us back. The day he arrived, there was more fighting and Father decided that it wasn't safe for the family to go back to Russia yet, so he hid his uniform in a secret closet, and we all stayed in Vilna.

After the revolution, conditions were terrible; it was practically famine—we really had very little to eat. Mother used to make all kinds of things out of potato peels in order to feed us somehow. We would get one lump of terrible bread for the whole family, full of straw and I don't know what else, and we bought a special pharmacist's scale to make sure it would be divided absolutely evenly. Even when we were hungry, Nadia was always the one who had the energy—she would disappear for a couple of hours and come back with a loaf of proper bread she had managed to find someplace. When we bought a large bottle of five hundred saccharine tablets (sugar simply didn't exist), Nadia would sit down on the steps and count out every one to make sure nobody had cheated us. We called her Saccharinchik after that...

Newta, 1985

Nadia soon was playing in the pit of the Gelios Theatre in Vilna, and Abrasha (Alexander) Schneider was in that little orchestra too. Grigory Katz, the owner of the theatre, was a married man, but he was very much in love with Nadia, and we were happy because he let us in to see all the movies free. The memory I've retained is of the spontaneous, nonsensical fun they had. I'll never forget Nadia, with her wonderful sense of humor, playing with one hand while she was eating a sandwich with the other.

Clara, 1984

Today they work for two, three years to make music for a movie; we had to look at the screen and make the music according to what was happening right then. We had on our stands music for crying, for killing, for anything, and we had to be able to improvise something immediately.

Alexander Schneider, 1984

Grigory Katz

Eventually, Father decided to leave Russia altogether. We went out of Vilna on a horse-drawn wagon, traveling all night to cross some sort of border. Of course, nobody asked the children: we were terribly unhappy about the whole thing. We begged and we screamed and cried and everything under the sun. How could it be possible that Nadia wouldn't go back to the Conservatory? What about our friends? Our life? Where in the world were we going?

Newta, 1985

At first, Father didn't know where we would settle, or how, but, even after he seriously started thinking about going to America, we kept moving from one country to another—wherever we could get visas. And my two sisters gave concerts all over the place, in Warsaw, in Danzig, in Memel...

Newta, 1985

At one concert, still in one of the small towns near Vilna, poor Nadia had to play on an impossible instrument which was tuned exactly half-a-tone low. Now both of us were blessed—or cursed—with absolute pitch, and for me to have to play everything down half a tone would have been a tragedy. Remember, I was only eight years old. Anyway, what did Nadia do? She said "Don't worry, play it exactly as it's written," and right on the spot she transposed all the music to the proper key. It was the most beautiful thing a big sister could have done for me.

Clara, 1984

It was a long journey. On the way we had to cross many borders illegally, and we traveled without any knowledge of whether or not we would ever reach a country from which we could get a visa to the U.S. Eventually we did, though, in 1922.

NR, 1974

It was a long, terrible voyage to America. Everybody was sick, particularly Mother and Father, and of course we traveled steerage or whatever they called it, all of us in one cabin, with bunks.

But we were young, after all. I remember Nadia wore a big hat and an imitation fur coat she got in Berlin, and I was wearing Father's coat from his Army uniform. Nadia was always a flirt (I was a little less so), so we met a number of boys on the ship and spent a lot of time with them.

And at the end of it was Ellis Island.

Newta, 1985

A New Life in America

I. J. (Sasha) Sherman

We met a journalist in Europe, and when he found out we were going to America, he said "I have a friend there, the former director of the Moscow Narodny (People's) Bank, I.J. Sherman. I'll write and ask him to meet you at the boat." That's precisely what happened, and Sasha was instrumental in getting us quickly out of Ellis Island.

NR, 1974

Newta, Sasha, Clara, and NR

From Ellis Island we went straight to Gardner, Massachusetts, where another uncle lived, and we sat there in Gardner for almost a month, which we needed because we were all exhausted and underfed. Then came a letter from Sasha to Father, saying that if he wanted to do anything with the children, musically speaking, he should bring them to New York. Father needed a job too—eventually he found one washing bottles in some chemical laboratory—and so the whole family went to New York, except me. I was extraneous, so they left me behind in Gardner for another few weeks.

Meanwhile, Sasha hired a little hall to introduce Nadia to some of his rather well-to-do friends, and collected quite a sum of money, enough for us to rent an apartment (on 7th Avenue at about 117th Street). He also got Clara a scholarship to the Ethical Culture School, and gave me my first job. I was all of 16 and didn't know a word of English, but Sasha was the head of the Soviet trade organization in New York, so he put me in front of a Russian typewriter and told me to learn how to use it. I actually was earning $15 a week...

Newta, 1985

To my dear Nadja with the wish that she may play the above in a few years like Hofmann.

Sept. 1922 Alex Lambert

Congratulations & great hopes

Alma Gluck Zimbalist

November 1/22

It was a great pleasure to listen to your playing on November 1st

Franz Kneisel 1922.

With all best wishes,

from Mischa Levitzki

Nov. 1, 1922

To Miss Reisenberg with best wishes and admiration for her beautiful playing.

Sincerely

Efrem Zimbalist

May 1922.

Sasha was of tremendous help in introducing us to different people, including Alexander Lambert, with whom I eventually studied.

NR, 1974

He also set up my first informal concerts in America—not debuts, but simply private evenings to earn some money. At that time it was fashionable for young artists to play at rich homes, and he arranged about ten or eleven recitals for me at $100 each, which was a very big sum then.

NR, 1974

Gradually, we became very friendly...
NR, 1974

NADIA
14th July 1922
In your heart you guard a fragile vessel,
With God's ember bit-by-bit starting to glow,
But if you let this tiny spark ignite your soul,
It will burst aflame and burn with sacred fire.
New York. I. Sherman
(For NR's first birthday in the U.S.A.)

Наде

14-е Июля 1922 г.

В душе своей сосуд ты хрупкий бережешь,
В нем искра божия чуть-чуть теплится,
Но если душу ты той искрою зажжешь,
Она огнем священным загорится.

Нью Йорк И. Шерман

The City Symphony Orchestra

DIRK FOCH
Conductor

Fourth "POP" Concert

Century Theatre, Sunday Afternoon, Dec. 17th

Soloist - NADIA REISENBERG - Pianist

Tschaikowsky....Symphony No. 6 (Pathetique) in B Minor, Op. 74

I—Adagio; Allegro non troppo.
II—Allegro con grazia.
III—Allegro molto vivace.
IV—Adagio lamentoso.

INTERMISSION

Paderewski.................Polish Fantasy, for Piano and Orchestra
Rimsky-Korsakow.........."Hymn to the Sun", from "Le Coq d'or"
Strauss, Johann...........................Waltz, "Blue Danube"

Tickets: Orchestra and Dress Circle $1.00, First Balcony 75c., and 50c., Second Balcony 25c., Box Seats $2.00.
On sale all week at Century Theatre Box Office.

Town Hall.........Wednesday, Dec. 20, at 3:00
Carnegie Hall..........Thursday, Dec. 21, at 8:30

Soloist - ERIKA MORINI - Violinist

Debussy..."Iberia"
Franck.....................Symphonic Poem "Chasseur Maudit"

INTERMISSION

Brahms..............Concerto in D Major, for Violin and Orchestra

I-Allegro non troppo.
II-Adagio.
III-Allegro giocoso, ma non troppo vivace.

Liszt.................................Hungarian Rhapsody No. 2

Tickets: Parquet $1.50, Dress Circle $1.25, $1.00
Balcony 75c, 50c, 25c; Boxes $20.00 and $16.00. (Tax Exempt.)
Now Selling at both Box Offices.

THE STEINWAY is the Official Piano of the City Symphony Orchestra.

Arthur J. Gaines, Manager.

Suite 921, 17 East 42nd Street. Telephone: Murray Hill 3305

CENTURY THEATRE

DIRECTION................MESSRS. LEE and J. J. SHUBERT

NOTICE: This Theatre, with every seat occupied, can be emptied in less than three minutes. Choose NOW the Exit nearest to your seat, and in case of fire walk (do not run) to that Exit.

THOMAS J. DRENNAN, Fire Commissioner.

SUNDAY AFTERNOON, DECEMBER 17, 1922, at 3:00

THE CITY SYMPHONY ORCHESTRA

DIRK FOCH, Conductor

"Dedicated to the Service of the People of New York"

Fourth "POP" Concert

Soloist—NADIA REISENBERG—Pianist

PROGRAM

Intermission

POLISH FANTASY FOR PIANO AND ORCHESTRA
OPUS 19........................Ignace Jan Paderewski
(Born 1860)

Soloist—NADIA REISENBERG—Pianist

While Paderewski is best known as a piano virtuoso, he has composed a large quantity of music. Unfortunately his reputation as a composer rests chiefly on the fame of one of his lesser works, the Minuet in G. But he has written music in most of the larger forms. His opera "Manru" was produced at the Metropolitan Opera House, he has created songs and chamber music, and for piano with orchestra he has produced three important compositions, Variations and Fugue, the Concerto in A minor and the Polish Fantasy which is being played at this concert.

Musical Ledger, Chicago, 12/21/22

PADEREWSKI LENDS INTEREST TO CITY SYMPHONY CONCERTS

Three concerts by the City Symphony Orchestra, Dirk Foch, conductor, were given. The program of Monday night at Carnegie Hall was repeated Tuesday afternoon in Town Hall, while the third was given at Century Theater on Sunday afternoon. The programs of the early week consisted of Tschaikowsky's "Pathetic" symphony and the prelude and "Love-Death" from "Tristan and Isolde" (Wagner). The audiences gained much pleasure from the growing homogeneity of the orchestra and Mr. Foch's readings. He still has a serious obstacle to overcome as he drags all slow movements and his presentations would gain mightily if he would heed this detail.

The greatest interest centered upon the Sunday afternoon concert when Paderewski was present to hear his own "Polish Fantasie" played by a Russian girl, Miss Nadia Reisenberg, at her American debut. Following her studies in Russia this young musician came to Alexander Lambert who found unusual material, and this eminent pianist and pedagogue never fails in the brilliant debuts which he is able to secure for his pupils.

It was a stroke of genius to have her play this work and coming as it did at the moment of the assassination of the president of Poland it carried new thrills for many including the composer who kept his word to be present notwithstanding his unhappiness over the tribulations of his country. It is a beautiful work and it received a fine performance by the soloist and orchestra. The applause was deafening at its conclusion, and Paderewski arose to acknowledge his share at which the audience arose to do him honor. The orchestra repeated Tschaikowsky's symphony and added the "Hymn to the Sun" from Rimsky-Korsakoff's "Coq d'or" and the Johann Strauss "Blue Danube" waltzes.

* * *

New York Herald, 12/18/22

Paderewski Hears Russian Girl Make Debut in His Music

Miss Reisenberg, 17, Plays Brilliantly as Soloist With City Symphony.

Ignace Jan Paderewski paid his tribute to his afflicted country as well as to youth in art, when, with a heavy heart at the news of the assassination of Poland's new president, he kept his word and went to the City Symphony Orchestra's concert at the Century Theater yesterday afternoon, where a young Russian refugee pianist, Miss Nadia Reisenberg made her American debut as the soloist in his own "Polish Fantasia" for piano and orchestra.

The performance of Mr. Paderewski's composition happened most timely. Fraught with Polish themes and color, the fantasy is a fine work of character and melodious beauty. Miss Reisenberg, who is 17 years old, first studied in Petrograd and later in this country one year with Lambert. Mr. Paderewski has shown interest in her talent and some time ago he said he would attend her debut of yesterday. Among the other prominent pianists who have marked her piano gifts, are Rachmaninoff and Hofmann.

The "Polish Fantasy" was well performed with no little credit going to the orchestra. The young player showed uncommon musical insight in her reading of the piano solo, giving it with the abandon and brilliance often found in a mature concert soloist. She was warmly applauded and in response to calls for the composer, Mr. Paderewski rose from his seat in a box to the left of the stage and acknowledged the applause while the audience rose in turn. A woman in the audience rushed frantically down the aisle, waving her program, and by her lively applause seemed to be exhorting all to continue the ovation to Mr. Paderewski.

Pani Nadi Reisenberg
z serdecznem podziękowaniem za bardzo piękne wykonanie mojej Fantazji Polskiej
I J Paderewski
28/XII 1922

To Nadia Reisenberg, heartiest thanks for a beautiful performance of my Polish Fantasy
Ignace Paderewski

PIANO RECITAL

BY

MISS NADIA REISENBERG

WYKEHAM RISE

Washington, Connecticut

MARCH 13

TUESDAY, ~~FEBRUARY 27~~, 1923

AT 3 P. M.

Highly Talented Nadia Reisenberg, in remembrance of her time spent within the walls of the Petersburg Conservatory. With heartfelt devotion.

Alexander Glazunov

Programme

I

a. Pastoral varie — Mozart

b. Prelude and Fugue A minor — Bach-Liszt

II

a. Nocturne — Chopin

b. Etude — Lambert

c. Valse — Glazounow

d. Theme and Variations — Glazounow

Programme

III

Polish Fantasie for Piano and Orchestra — Paderewski

Orchestra part played on a second piano by Mr. Lambert

Lambert called and said that I have to play somewhere out of town in some school. We'll go there in the morning and return late at night. We'll do the Paderewski—he'll be at the second piano.

NR diary, 1923

To Miss Nadia Reisenberg
~~In~~ admiration of her
great talent and ~~her~~ beautiful
playing
from
Mischa Elman
New York Nov. 23/1923

To Miss Nadia Reisenberg, the young
and talented pianiste, in remembrance
of her splendid rendition of Albeniz'
"Triana" (in the original discordant version).
With best wishes for her great
success. Leopold Godowsky
New York, Oct. 2nd 1923

I don't remember too much about my recital debut except that it was very well received and that I had wonderful reviews. More engagements came pretty soon afterwards, and it all started.

NR, 1974

New York Evening Journal, 2/7/24

An Excellent Young Pianist.

Nadia Reisenberg, a young pianist who played here a year ago or so with the defunct City Symphony Orchestra, gave a recital last evening in Aeolian Hall—her first appearance on her own. She made an unusually favorable impression, not only on her large and rather friendly audience, but on the more detached critical observer as well.

The young woman has gained a good deal in resourcefulness and poise since she played here before. She has acquired a more than respectable technic and she gave one the impression last evening that she knew what she was about in the interpretation of the music she had in hand.

Her programme was an unusual one and showed that she had given considerable thought to putting it together. But she had given even more thought to the necessities of expression in the varying types of music on it. She played one of the organ preludes and fugues of Bach, for instance, with a fine clarity that was never dry; and she was able to shift her point of view immediately to the quite different requirements of a piece like Mozart's "Pastorale varieo." This she played with a delightful delicacy—really a thorough appreciation of its spirit. And so she shifted her ground from one piece to another, finding the right mood for each as she went along. Altogether, a most promising young pianist.

New York Herald, 2/7/24

NADIA REISENBERG PLAYS.

Young Pianist Wins Approval of Large Audience.

Miss Nadia Reisenberg, pianist, gave her first recital here last night in Aeolian Hall. Born in Petrograd, she later was a refugee in Poland, and, reaching here two years ago, became a pupil of Alexander Lambert. Last season she made her debut as soloist with the City Symphony. Now, it is said, she is to play here and on tour with the New York Symphony.

She was heard last night by a large audience and won enthusiastic approval. Her performance again marked her as a gifted player of brilliance and finish far beyond her years, now 19. Her style tends to that of the virtuoso rather than to that of a player of poetic introspection.

In the Bach-Liszt prelude and fugue, in A minor, she played with admirable freedom and poise, and in the theme and variations of Glazounov she tossed off technical difficulties with ease and fluency. A little more time spent in study and experience will no doubt mellow and add finish to her art. She is already a very interesting player.

New York Evening World, 2/7/24

Nadia Reisenberg, a young Russian pianist, unusually gifted, gave her first New York recital in the evening in Aeolian Hall. She may be recalled by some as a soloist last season with the City Symphony Orchestra playing Paderewski's Polish Fantasie with the composer looking on. Born in Petrograd, Miss Reisenberg arrived here, a refugee, at the age of seventeen, having started her musical studies when she was twelve. The young lady's outstanding asset at present is her genuine musical feeling. An essential any teacher is glad to light upon. She has a gentle touch, a good tone, a strong feeling for rhythm, a nice sense of proportion, an attractive playing style, understands the use of the pedal and already there is an individual stamp upon her playing. Her runs and swifter passage work were smooth and musical. Altogether a successful occasion for the young pianist.

New York Times, 2/7/24

Nadia Reisenberg, Pianist, Pleases.

Nadia Reisenberg, a young Russian pianist of evidently rare musical nature, won the favor of her first New York recital audience last evening in Aeolian Hall. A student at 12 years in her native Petrograd, she had reached America at 17 as a war refugee and had been once heard last year with the City Symphony, playing Paderewski's "Polish Fantasy." The dark, slender girl gave a fresh vividness to last night's music in a theme and variation of crystalline charm by Glazounov and lesser pieces by Medtner and Scriabin. There was a preface of elder classics, including the Rameau-Godowsky "Tambourin," and a final group from Liszt, Debussy and Albeniz.

New York American, 2/7/24

* * *

A PIANIST of unusual musicianship is Nadia Reisenberg, whose evening recital at Aeolian Hall gave her a chance to demonstrate her gifts in the Bach-Liszt A minor prelude and fugue, Glazounow's theme and variations, and shorter compositions by Mozart, Scarlatti, Chopin Rameau-Godowsky, Medtner, Scriabine, Liszt, Debussy, and Albeniz.

Miss Reisenberg approaches her tasks in a serious, straightforward fashion and conquers them convincingly. Her interpretative processes have particular interest, for her responsive temperament never permits her to fall into mere scholarliness. She has, too, ample strength, agility and brilliancy of technique. She is altogether a most worth while young piano exponent.

* * *

Musical America, 2/16/24

Nadia Reisenberg's Recital

Nadia Reisenberg, a young Russian pianist, who made her début here last year with the City Symphony, gave her first recital at Aeolian Hall on Wednesday evening, Feb. 6. She impressed her audience with the vivid quality and facility of her playing. In the Bach-Liszt Prelude and Fugue in A Minor she displayed a virtuoso style, and in the Glazounoff Thème et Variations she seemed oblivious to the technical difficulties. In the rest of her program, which ranged from Scarlatti to Scriabin and from Liszt to Debussy and Albeñiz, she combined an almost crystalline finish with youthful poise and charm. One of the largest début audiences of the season was perhaps as delighted by her slim dark beauty as by the freshness of her style. H. M.

Musical Digest, 2/19/24

Nadia Reisenberg

Nadia Reisenberg, a young Russian pianiste, made her New York debut on Wednesday evening a week ago. She played music by Bach, Glazounov, Scriabin, Rameau-Godowsky, Liszt, Debussy and Albeniz. The reviewers were unanimous in commending the new pianiste.

The Times describes her as "a pianiste of evidently rare musical nature." According to the Herald, "she played with admirable freedom and poise. She tossed off technical difficulties with ease and fluency." Paul Morris of the Telegram discovers "a light rhythmic touch; she makes her piano sing and she plays with the utmost clarity and precision." In the words of the Sun, "she did some unusually good things to prove that she knew what she was playing and how to play it."

New York Tribune, 12/7/24

At Aeolian Hall, Nadia Reisenberg, a young Russian pianist, who had made a very favcrable impression last season at a Century Theater concert of the City Symphony Orchestra, gave a recital beginning with Liszt's version of the Bach A minor Prelude and Fugue, followed by Mozart and Scarlatti. Miss Reisenberg repeated the impression of vigor and technical brilliance made last season; her playing combining clearness with capacity for high speed. The Bach-Liszt number, well played from the mechanical standpoint, was rather formal, nor was Mozart's "Pastorale Variee" particularly varied, but the third group, of Chopin, Medtner and Scriabin, showed brighter coloring and the impression of brilliance noted in Miss Reisenberg's earlier appearance. Glazounoff's Theme and Variations was the second group, while Liszt, Debussy and Albeniz ended the program of a concert very well attended and much applauded.

The Sun and The Globe, 2/7/24

In Aeolian Hall, meanwhile, Nadia Reisenberg, a young pianist who made her debut as soloist with the late City Symphony last season, gave her first local recital. She did some unusually good things to prove that she knew what she was playing and how to play it. There was keen constructiveness in her interpretations, with a lively style, clean touch and capable technique behind them

To beloved Nadia, in remembrance
Alex Lambert New York, 1922

Aeolian Hall

THE SYMPHONY SOCIETY OF NEW YORK

1923—*Season*—1924

New York Symphony Orchestra

WALTER DAMROSCH, Conductor

Fifteenth Sunday Afternoon Subscription Concert

March 23d, at 3 o'clock

Soloists

Miss NADIA REISENBERG

Mr. MARCEL GRANDJANY

PROGRAM

I. Symphony in D minor....................*Franck*

I. Lento. Allegro non troppo

II. Allegretto

III. Finale. Allegro non troppo

Program continued on next page

PROGRAM CONTINUED

II. Variations Plaisantes, for Harp with Orchestra *Roger-Ducasse*

Mr. Marcel Grandjany

III. Pavane ....*Fauré*

IV. Concerto, for Piano with Orchestra......*Rimsky-Korsakoff*

Miss Nadia Reisenberg

V. Marche Américaine ....*Widor*

(New; first time)

(*The Steinway is the Official Piano of the New York Symphony Orchestra*)

George Engles, Manager

New York Sun, 3/24/24

Mr. Damrosch Gives Varied Program

An unusually varied program was offered by Mr. Damrosch for the New York Symphony's concert in Aeolian Hall yesterday afternoon. Interest centered in the performance of a long neglected concerto by the distinguished neo-Russian composer Rimsky-Korsakoff. The assisting artist was Miss Nadia Reisenberg, a gifted young pianist who was heard with the City Symphony Orchestra last season and who gave her first recital on February 6 of this year. Available records fail to disclose any previous performance of this concerto in New York, and indeed this fact is not surprising. The concerto is an early work of the great Russian, who composed it in 1882. It is brief, spirited and neither very imposing in musical content nor impressive in structure. However, it proved a vividly colorful and superficially brilliant piece which afforded many a kindly opportunity for Miss Reisenberg to shine forth in radiant hues. Rather nervous at first, she rapidly gained poise and assurance. Her expressive shading and sensitive touch won her enthusiastic and well deserved applause.

Marcel Grandjany, the eminent French harpist, gave much pleasure with the "Variations Plaisantes" of Roger-Du-

New York American, 3/24/24

Conductor—Walter Damrosch.

Soloists — Nadia Reisenberg, piano; Marcel Grandjany, harp.

Performance — Arrived during "Variations Plaisantes." (Pleasant variations). They are pleasant in the sense that they are not unpleasant. Modern writing, sophisticated rather than seductive. Grandjany a splendid harpist of unfailing taste and refined musicianship. Faure's "Pavane" placidly soothing. Rimsky-Korsakoff concerto (amazingly reminiscent of Liszt's in E flat) a mass of glittering generalities. Agreeably short. Played by Miss Reisenberg with a degree of devotion, dash and technical command worthy of a better cause. A young lady of unusual and impressive talent. Widor's march more Hungarian than American in coloring, more paprika than pep. Lively enough, to be sure. Damrosch informed audience Widor aged sixty-eight when he wrote this joyous music. "The unquenchable youthfulness of France."

Among those present—Leopold Godowsky, Josef Hofmann, Alexander Lambert, all in one box. Suspect Lambert of pedagogical, patriarchal interest in pianist.

APEDA
N.Y.

Zu Nadia Reisenberg
en admiration de Votre
Talent pianistica

V. de Pachmann

15.5.24.

To Nadia Reisenberg
In admiration of your pianistic talent.
Vladimir de Pachmann

Family Interlude I

Mr. and Mrs. Aaron J. Reisenberg
announce the marriage of their daughter
Nadia
to
Mr. Isaac J. Sherman
on Tuesday, the twenty fourth of June
One thousand, nine hundred and twenty-four
City of New York

Top row: Aaron, Rachel, NR, and Sasha. Bottom: Newta, Clara, and Boris Elperin

H ________________ LINE
(Name of Line)

CERTIFICATE OF IDENTIFICATION
To Facilitate Re-Admission to the United States

(To be retained with the passport by the passenger and presented to the Steamship Company's office in Europe as early as possible, in order to eliminate unnecessary delays on the return trip.)

STATE OF New York } ss.
COUNTY OF New York

I Nadzieja Sherman being duly sworn, depose and say:

That I am temporarily leaving the United States, sailing on (date) July 12th 1924 from the port of New York to the port of Southampton aboard the S. S. Homeric of the White Star Line, and hold first class eastbound ticket No. J 72648, and that I am being accompanied by the following members of my family:

Name	Country of Birth
Isaac Sherman (Husband)	U.S. Citizen

That my temporary absence abroad is for the purpose of visiting relatives

and that I intend to return to my permanent home in the United States within about six months.

That I am 20 years old, am (married or single) married and was born in (country) Poland

That I now reside at No. 1971 Seventh Ave in the City or Town of New York State of New York and have lived in the United States continuously for 2½ years.

Subscribed and sworn to before me this 9th day of July 1924

Hiram Grover
Notary Public
NOTARY PUBLIC NEW YORK COUNTY
CLERK'S NO. 427, REG. NO. 6011
COMMISSION EXPIRES MARCH 30, 1926

Nadzieja Sherman
Signature of Passenger

PRINTED IN THE U. S. A.

They went to Russia on their honeymoon, and when they came back, Nadia was already pregnant with Alex, and Clara was counting the months and the days...

Newta, 1985

□□□ Р. С. Ф. С. Р. □□□

НАРОДНЫЙ КОМИССАРИАТ ПО ВНУТРЕННИМ ДЕЛАМ.

Взыскано
[illegible] 5 руб. [illegible] коп.
„[illegible]“ [illegible] 192[illegible] г.
гор. [illegible]

Виза № [illegible]

Действительна по „[illegible]“ [illegible] 192[illegible] г.

Владельцу [illegible]

от [illegible] № [illegible], к которому прикреплена настоящая виза гр. [illegible]

Разрешен выезд за пределы С. С. С. Р. в [illegible] через Пограничный Пункт [illegible]

По уполномочию Народного Комиссара
по Внутренним Делам РСФСР

Зав. Инотделением Губотуправ

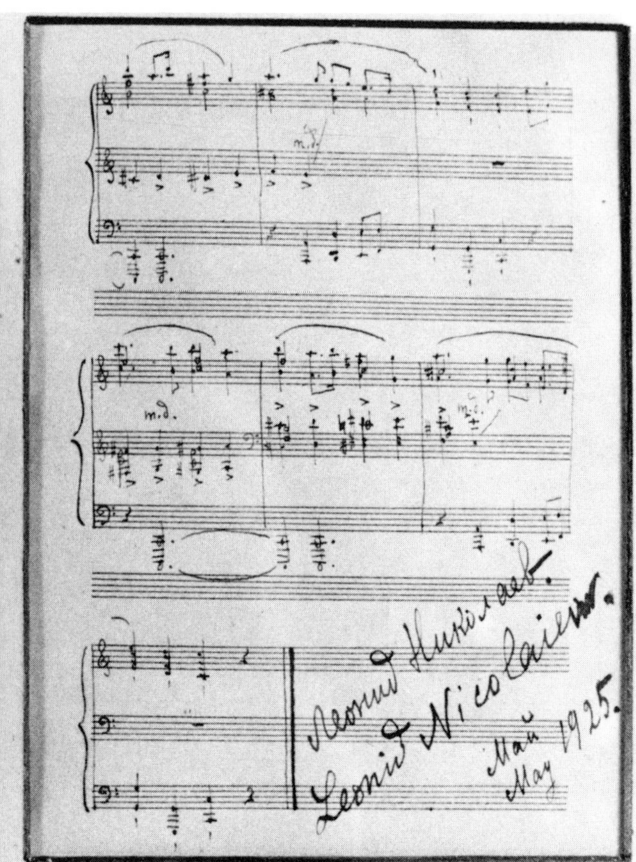

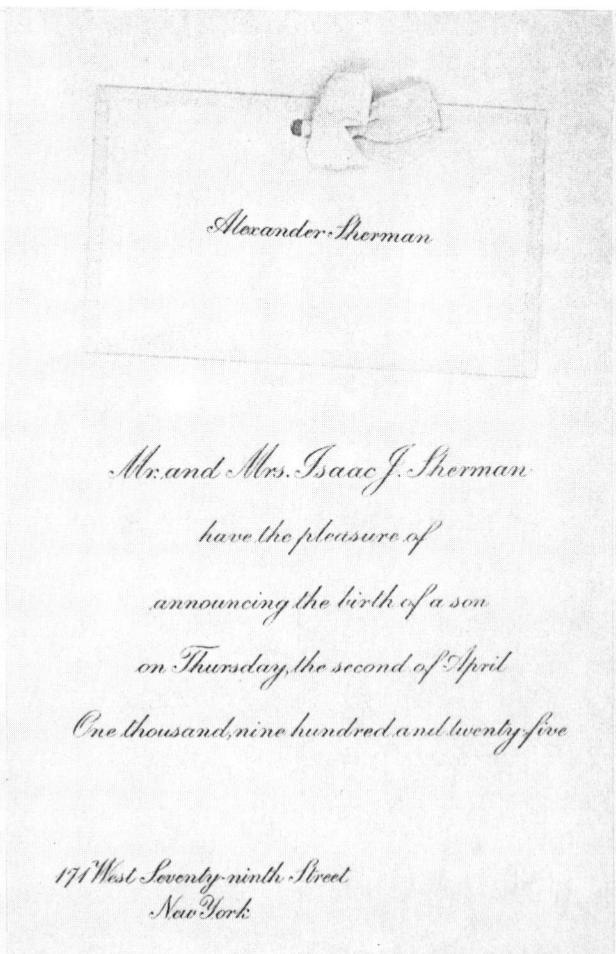
Alexander Sherman

Mr. and Mrs. Isaac J. Sherman
have the pleasure of
announcing the birth of a son
on Thursday, the second of April
One thousand, nine hundred and twenty-five

171 West Seventy-ninth Street
New York

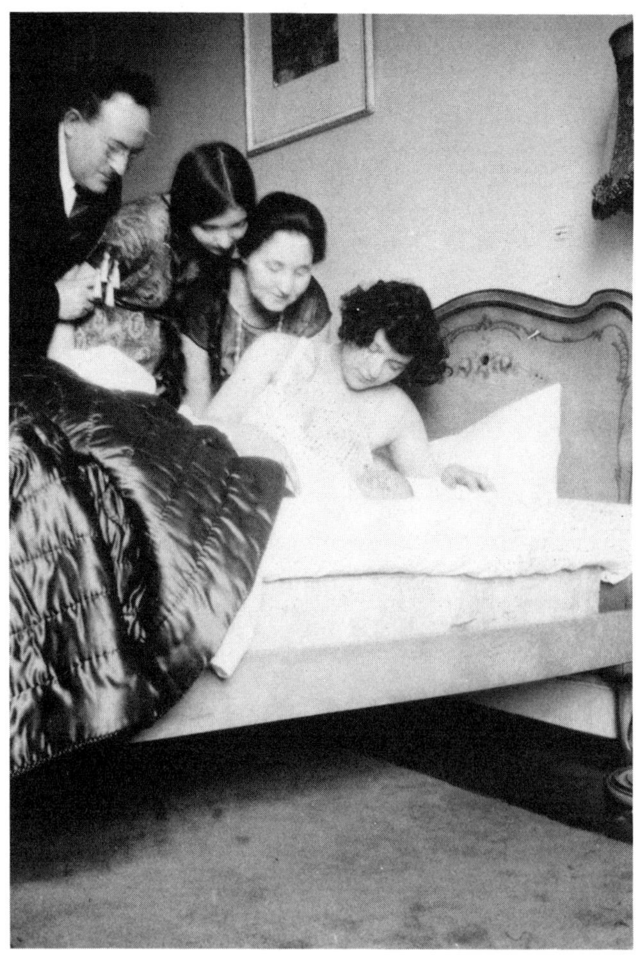

NR and first son, Alex

Top: Rachel, Aaron, and NR.
Bottom: Alex and Sasha

WEDNESDAY, MARCH 24, 1926.

Music : Olga Samaroff's Daily Column

Nadia Reisenberg—Stokowski

NADIA REISENBERG, who, I am told is generally considered one of the most gifted of the younger pianists now making a place for themselves in New York concert life, gave a recital at Aeolian Hall yesterday afternoon. Hearing her for the first time, I am inclined to agree with the prevailing opinion. She is a pupil of Alexander Lambert and shows the kind of sound, brilliant piano-playing he is so well able to develop, plus many natural qualities which cannot be taught.

Olga Samaroff

Her program opened with the D minor and F minor (not C minor, as printed in the program) preludes and fugues from the second book of the Well-tempered Clavichord of Bach. While Miss Reisenberg's Bach playing did not seem to me as unusual as things she did later in the program, it was good and showed solid musicianship.

The Scarlatti-Tausig Pastorale and Capriccio were so well done that I think if Miss Reisenberg were playing this work behind a screen in competition with many pianists of high standing it would have been difficult for the latter to excel her in style, clarity and delicacy.

It seemed a pity to waste good playing on such an empty and uninteresting composition as the "Rigaudon" of Raff, which closed the first part of her program. Its only possible merit in my opinion might be a display of certain pianistic qualites, but our rich piano literature contains innumerable opportunities for such display in connection with real musical value.

Miss Reisenberg's playing of the Schumann G minor sonata had many fine qualities. I did not always agree with her pedalling and it seemed to me that she might develop more luscious beauty of tone for such lyric passages as the melody of the andantino, but there was a fine sweep and youthful ardor in her conception of the work as a whole. In the Glazounow Theme and Variations she did her best playing of the afternoon. Her auditors were highly attentive and very responsive.

Detroit Evening Times, 11/26/26

Clever Girl Pianist Makes Hit at Sunday Pop

11/22/26

Miss Nadia Reisenberg's spirited performance of Paderewski's "Polish Fantasy," and Victor Kolar's reading of Charpentier's "Impression of Italy" and the introduction to the third act of "Lohengrin" were the high lights of the sixth Sunday afternoon programme of the Detroit Symphony Orchestra in Orchestra Hall yesterday afternoon.

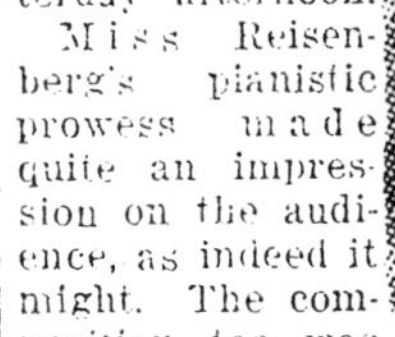
Nadia Reisenberg

Miss Reisenberg's pianistic prowess made quite an impression on the audience, as indeed it might. The composition, too, was one calculated to arouse enthusiasm, for it is mostly a dashing melange of Polish folk songs and dances, sometimes with a contrasting interlude of romance, but for the most part a swift-moving panorama. Written more than 30 years ago when the young Paderewski must have been first flaming into spectacular fame in this country, it abounds in pianist effects which are showy and, I suspect, not exactly easy to execute. But Miss Reisenberg handled them easily, with the staunch support of [illegible] ...ded orchestra.

Dayton Journal, 1/28/27

RUSSIAN PIANIST CHARMS AUDIENCE AT CLUB CONCERT

Naida Reisenberg Presents Second of Matinee Musicals At Miami Hotel.

Beauty of person, charm of personality and a rare ability as a pianist are attributes of Naida Reisenberg, Russian pianist, who presented yesterday morning in the Miami hotel ballroom the second of a series of matinee musicals which is being given under the auspices of the Dayton Women's Music club.

The pianist has a depth of tone, her technique is clear-cut, her fingers are marvelously agile and her wrists very flexible. There is a color to her playing which speaks at once of culture and of spirituality.

The program which Miss Reisenberg presented was not an easy one. It opened with a brilliant Scarlatti "Capriccio," arranged by Tausig. Next came a Godowsky arrangement of Bach's "Suite No. 2 in D Minor." It speaks well for Miss Reisenberg's artistic attainment to say that she played these difficult selections in such a manner that they seemed easy.

The pianist presented two of Gounod's selections. The first of these was "Theme and Variations;" the second, a waltz which she gave as a final encore.

Liszt's lovely "Etude de Concert, D Flat Major," Paganini-Schumann "Caprice in E Major" and Chopin's "Valse in A Flat Major" were among the most delightful selections on the program. As an encore to these, she played a rarely heard Chopin nocturne.

"Islamey," by Balskirew, the last programmed selection, is one of the most difficult numbers in the entire repertory of piano music. It calls for superlative skill in its presentation. The pianist had the skill.

Boston Traveler, 12/14/26

PIANIST MAKES BOSTON DEBUT

Nadia Reisenberg Is Soloist with the Symphony Orchestra

A young Russian pianist, Nadia Reisenberg, made her first Boston appearance with the Boston Symphony Orchestra last evening, when Sergo Koussevitzky conducted the second of the Monday evening series of concerts at Symphony Hall.

Miss Reisenberg has studied in New York, where she has made her home since the Russian revolution, and has played there both in recital and as soloist with the New York Symphony Orchestra.

Last night she was heard in Liszt's second pianoforte concerto, in A major. This taxing work exhibited her as a pianist of power and brilliance, equal to all the demands of the occasion Perhaps because of her youth and chic appearance, the audience forgot to credit Liszt, Koussevitzky or the orchestra for any share whatsoever in the performance, and lavished all its vigorous hand-clapping on the deserving young soloist, when there should have been enough glory for all.

The program began with the fourth symphony of Johannes Brahms, in a glowing performance that roused the audience to a high pitch of enthusiasm. The concerto was heard after the intermission, and was followed by Berlioz's "Roman Carnival" overture, which completed the concert with a grand flourish.

Boston Evening Transcript, n.d.

Miss Reisenberg possesses all the accomplishments of the soloist who would appear successfully with an orchestra such as the Boston Symphony. Her technic can meet any demands—in facility, as an agent of color, in producing volume of sound. Of equal excellence is the tonal quality of her melodies. To it all she brings discerning musicianship. With Liszt she can languish and with him she can strike fire. And above all she plays with a will, a will of her own. Not altogether is she satisfied to take tempos and moods from Mr. Koussevitzky. With head and shoulders tracing her rhythms in the air, time and again she sets her own tempos, plans her own interpretations, which conductor and orchestra follow. And to no disadvantage does she do this. For hers is an understanding admirably equipped thus to direct the course of much of the music she plays. One would welcome Miss Reisenberg in a recital of her own.

Musical Courier, 4/1/26

Nadia Reisenberg

On March 23, Nadia Reisenberg delighted a representative Aeolian Hall audience with her splendid pianism and genuine artistic abilities with nimble fingers, admirable technic, and a finished style and true conception of her numbers, she gave a program that set off her brilliant pianism to distinct advantage. Her presentation of Schumann's sonata in G minor was perhaps the outstanding number of her program, and she gave an interpretation that brought forth the enthusiastic plaudits of the audience. Three preludes and fugues from the Well Tempered Clavicord by Bach, and selections of Scarlatti-Tausig and Raff, were rendered with a verve and spirit that made them most entrancing. She offered a clear-cut rendition of Glazounoff's Theme et Variations, and a group of Chopin, Liszt and Scriabin selections rounded out her program. All in all, Miss Reisenberg made a genuine impression on those who heard her, and merited in every way the warm reception she was accorded. It might be added that she also received excellent commendations from the press.

Chicago Evening American, n.d.

Naida Reisenberg—pianist at the Blackstone Theater—new name—debut here. A second Guiomar Novaes—an unfair comparison, for Nadia Reisenberg is a power and a personality in her own right. All I heard was the Glazounow "Theme and Variations"—enough to pronounce her a wonderful, refreshing, genuine, ripened talent.

* * *

Chicago Daily Journal, n.d.

Making Personality Count

Nadia Reisenberg, making her local debut in piano recital at the Blackstone theater yesterday afternoon, proved to be one of those rare persons with well-grounded and easily communicable enthusiasms of an individual sort. She played with conspicuous ability, for she has a genuine and resourceful craftsmanship, even though the portion of her program in which she was heard did not suggest she has yet found a unity or completeness of style for all the fine things she so engagingly seeks to express.

Chicago Daily News, 1/31/27

Nadia Reisenberg Triumphs.

Miss Kinsolving's presentation of Nadia Reisenberg to a Chicago audience yesterday at the Blackstone theater was an achievement on which she is to be congratulated. Miss Reisenberg is a Russian girl, loudly proclaimed by the musical elect, but new here. She will be wanted often hereafter, for she is a pianist of unusual distinction, with individuality of conception, a brilliant execution and exquisite finish.

The song recital at the Goodman theater by Leola Turner, soprano, was one of wide variety, including compositions of Scarlatti, Bach, Hugo Wolf and Mahler and folk songs of France and Italy. Miss Turner has a pleasant voice of considerable power and she uses it well.

New York Times, 1/5/27

NADIA REISENBERG HEARD.

Pianist Stirs Audience That at First Shows Coolness.

Nadia Reisenberg gave a piano recital at Aeolian Hall last night before an audience which appraised her somewhat coolly at first, but which became most enthusiastic by the end of the evening.

The perfection of a remarkably graceful technic and a thorough knowledge of the classic style were observable in the opening Bach Fugue, but in Schubert's Sonata in A major, which followed, the artist's playing was disappointing, except in the Allegro.

The Bach Suite for violoncello, transcribed and adapted for piano by Godowsky, was quite another story, and threw an illuminating light on the pianist's reserves of force. The prelude and Allemande were given masterfully and with splendid decision, quite in the Bach manner. The remaining movements, especially the final Gigue, were more in the language of Godowsky, and therefore more flowery.

The Liszt Etude de Concert had a brilliance, a sparkle, that galvanized the audience into demonstrations of pleasure. The Paganini-Schumann Caprice had its use as a pièce de concert.

Miss Reisenberg became astonishingly temperamental in Balakirev's "Islamey," which ended the concert. No one, having heard her play Bach with such exactitude and decorum, would have dreamed that she could become so flashingly barbaric and bacchantic. The audience applauded enthusiastically and recalled the artist to an encore.

Boston Evening Post, 12/17/26

SYMPHONY PRESENTS NEW ARTIST

Nadia Reisenberg Is a Pianist of Unusual Skill

BY WARREN STOREY SMITH

Last evening at the season's second Monday concert of the Symphony Orchestra a new pianist of no ordinary quality was introduced to Boston. Nadia Reisenberg is her name; Russia was her birthplace; New York is her present residence; and to her unusual pianistic accomplishments and musical gifts the kindly fates have added no small degree of personal charm.

LISZT'S A MAJOR CONCERTO

For her "vehicle," as the expression goes, Miss Reisenberg chose last evening Liszt's Concerto in A major, a piece that is not so much music, in the highest sense of the word, as an amazingly ingenious compilation of tonal tricks. But Miss Reisenberg, with Mr. Koussevitzky and the orchestra to give substantial aid, played this gilt and tinsel and pseudo-poetizing in a manner to blind the listener to its inherent emptiness.

First of all, Miss Reisenberg is mistress of the arts of musical rhetoric; her playing has the quality that compels attention. To judge by the ease with which she surmounted the difficulties that bestrew the pianist's path in this concerto, technique is for her a matter of second nature, while a lovelier tone than hers, and a more persuasive knack of moulding a melody, is not often to be met with. Nor must the general statement that the entire performance of this concerto last evening was brilliant and effective in the extreme go unenforced by the specific mention of Mr. Bedetti's beautiful playing of the few measures for solo 'cello in the slow division. It is hardly necessary to add that Miss Reisenberg was liberally applauded by the large audience.

15476

FEBRUARY 24, 1928 The American Hebrew

A Gifted Recitalist

Arduous Study Has Won Her Acclaim of Paderewski, Hoffman and Bodanzky

By CLAUDINE BROWN

NADIA REISENBERG
An accomplished pianiste whose keyboard technique has clarity, delicacy and ardor. She has appeared as soloist with the New York, Boston and Detroit Symphony Orchestras

AT her pretty apartment in West Eighty-seventh Street, Miss Nadia Reisenberg, the accomplished young Russian pianiste who, in private life, is Mrs. Isaac J. Sherman, greeted the interviewer in a cordial manner. Her appearance is girlish and it was something of a surprise to learn that she is not only married but is the mother of a two-and-a-half year old son, who is, she says, "the best thing she has ever done."

Miss Reisenberg was born in Vilna, Russia, and began to play at six years of age. At nine she entered the Imperial Conservatory of Petrograd where she received her first training under Leonid Nikolaiev, a teacher of high reputation; and she later studied here in New York with Alexander Lambert.

Before coming to America, Miss Reisenberg made a concert tour through Europe. She was accompanied by a younger sister who is talented violinist, now studying with Leopold Auer, and who will make her début next year.

In 1918-19 she left Petrograd and returned to Vilna. In order to escape the effects of the Revolution in the land of her birth she and this younger sister played in concerts for the benefit of orphan children and after that received permission to leave. This was accomplished through the influence of the wife of a high official interested in the orphanage. She tells tells a vivid story of the hardships they suffered and the extreme hunger. Though her family were people of means and lived in comfort, food was so scarce they were unable to buy it at any price. Her father was not permitted to accompany them at first, but later joined them, and the entire family is now in America. She tells how she chanced upon a friend in London, who learning that she was on her way to America, wrote to a friend of his over here requesting that he meet Miss Reisenberg when she landed. The pianist relates, with a twinkle in her eye, that his friend met her and is now her husband. Miss Reisenberg returned to Russia a few years ago on her honeymoon.

While studying with Lambert in America and after her first recital, her personality and brilliant technical accomplishments attracted the enthusiastic approval of such musical notables as Pederewski, Hoffman, Bodanzky and Damrosch. She has had subsequent engagements with the New tables as Paderewski, Hoffman, Boton and Detroit Symphony Orchestras and others, and given many recitals.

When she first came to New York she played at the Capitol Theatre because of the necessity of earning money. She found irksome the monotony of repeating her program twice daily for seven days.

Miss Reisenberg is much interested in modern music and its development —particularly the Russian composers. The ultra modern music she is not, as yet, ready to accept, but is interested also in its development.

She finds the American people most hospitable and cordial and says everywhere she has traveled people have been so interested and kind.

A glimpse about Miss Reisenberg's home shows her interest in other artistic fields than music. Many lovely paintings adorn her home and the writer had a peep at an unfinished portrait of the artist herself, being done by a famous Italian portraitist. She says she could not conceive of life without paintings around her. They mean so much both to her and her husband, and she deplored the lack of paintings in so many American homes. She is interested in books and has quite a good library and also a very fine musical collection.

Miss Reisenberg played last week as soloist with the Roxy Symphony Orchestra, augmented to 110 pieces, under the direction of Erno Rapee. Her program included the overture to Oberon and Grieg's "Peer Gynt."

50th Street and 7th Avenue, New York

S. L. Rothafel, "ROXY", Director

... FIFTEENTH ...

SYMPHONY CONCERT

SUNDAY, FEBRUARY 12th, 1928
AT 11:30 A. M.

ROXY SYMPHONY ORCHESTRA
110 Musicians

Conductors: Erno Rapee, Charles Previn and Joseph Littau
Assistant Conductor: Mischa Violin
Concertmasters: Henri Nosco and Josef Stopak
Solo cellist: Yascha Bunchuk

ERNO RAPEE, conducting

Soloist: *NADIA REISENBERG, Pianist*

---

sun. The second number, "Aase's Death," was written to accompany the death of a lonely mother lamenting for her son who has forsaken her. "Anitra's Dance," the third number of the Suite, is a lively Mazurka, representing Anitra, a lithe, supple Bedouin girl dancing with her maidens. The final number, "In the Hall of the Mountain King," is a fantastic piece picturing Peer Gynt in the Cave of the Gnomes.

4. *Rondo Capriccioso* .... *Felix Von Mendelssohn*

MISS REISENBERG

Mendelssohn wrote this composition in one of his lighter moods. It is colorful, vivacious, brilliant and thoroughly pianistic in style. Mendelssohn was a genius who died very early in life. One feels sure that great things would have evolved had he lived longer.

5. *Hungarian Rhapsody No. 2* .... *Franz Liszt*

The second Hungarian Rhapsody is the most popular of all of Franz Liszt's fifteen Rhapsodies. While originally, all of these works were written for piano, Liszt, aided by Franz Doppler, arranged six of them for orchestra. The thematic material for the Hungarian Rhapsodies was derived chiefly from gypsy melodies, many of which had never been recorded on paper. The editor of the Ditson edition of Liszt's Hungarian Rhapsodies has suggested that "tunes which before had served to amuse the motley crowd at the Czardas on the Puszta, have through Liszt been successfully introduced into legitimate music." "Most wonderful of all," he continues, "he has not hesitated to preserve all the drastic and coarse effects of the gypsy band without ever leaning toward vulgarity. To revel in sombre melodies seems to be one-half of the purpose of Hungarian music, and—in logical opposition—a frolicsome gaiety the other half."

The Sixteenth of the Series of the Symphony Concerts by the Roxy Symphony Orchestra, Erno Rapee Conducting, will take place next Sunday Morning, February 19th at 11:30.

Rarely do I experience the elation I find in playing chamber music. It's a conversation—you speak, I speak—and there is something very wonderful about doing things together. When I play a trio or a quartet, I don't only see my two lines, I see the entire score . . .

NR, 1973

The sense of give-and-take, of listening, of being aware of all the sounds around you lets you ripen into a better musician . . .

NR, 1974

When we play a concerto, it's our right to show off—we are the soloist, we are playing something brilliant, we are the main thing, our own way of doing things can be in front. Never in chamber music! This is the time when our own ego goes back. It's no longer *I*, it's *we*, it's *us*, it's the ensemble, it's the music . . .

NR, 1981

The Town Hall

SEASON 1929—1930

Wednesday Evening, October 23, at 8:30

FIRST SUBSCRIPTION CONCERT

The Stringwood Ensemble

JOSEF STOPAK, *First Violin*
MISCHA MUSCANTO, *Second Violin*
MICHAEL CORES, *Viola*
ABRAM BORODKIN, *'Cello*
SIMEON BELLISON, *Clarinet*
NADIA REISENBERG, *Pianist*

PROGRAM

I

Trio in E Flat (Kochel No. 498) .... *Mozart*

Andante

Menuetto

Allegretto

(For Clarinet, Viola and Piano)

Program Continued on Second Page Following

SERGE KOTLARSKY *First Violin* — SIMEON BELLISON *Clarinet* — ABRAM BORODKIN *'Cello*

MICHAEL CORES *Viola* — NADIA REISENBERG *Piano* — SAMUEL KUSKIN *Second Violin*

Technique, *per se,* doesn't exist—it's only a means of making music. You have to be free, to have enough equipment to allow you to deal with the most difficult things effortlessly. An important aspect of this is economy of motion. Many people look so busy at the keyboard—they move everything around, including themselves—but if they would sit quietly and just *listen* to the sound they produce, and use the minium of unnecessary motion, how much faster they could play,how much more accurately, how much more beautifully.

NR, 1977

The 1930s

Budapest

Josef Hofmann

Milutkiej i utalentowanej
Nadi Reisenberg
na pamiątkę
od
Józefa Hofmann
N.Y. 1922

Note Rachmaninoff signature in lower left corner

Yesterday heard Hofmann. I think he never played so beautifully before. The nearest to my heart was Scriabin. After the concert talked to Hofmann. Thanked him for Scriabin. Honestly, he played like a god.

NR, diary, 1923

By virtue of the power invested in them by the
Commonwealth of Pennsylvania
and upon the recommendation of the Director and the Faculty, the Officers and members of the Board of Directors of The Curtis Institute of Music hereby confer upon

Nadia Reisenberg

this

Diploma

in recognition of the honorable and satisfactory completion of her studies at this Institute

Given at Philadelphia, Pennsylvania, this twentieth day of May, 1935.

Josef Hofmann with his Summer Seminar group. From left: Shura Cherkassky, Nadia Reisenberg, Mr. Hofmann, Joseph Levine and William Harms.

TULSA

March 3rd
1933

My dear Nadia,
your letter interested me. I was glad to learn that you were pleased with yourself for you are a better judge than a silly critic. Console yourself. One of the western critics started his report about me as follows: "a head full of notes and two mighty hands that is Josef Hofmann". Good characteristic of — a musical blacksmith! Am well and pleased with my tour. Have been playing rather well. I do not know why. Weather was "poifect" all along and

the air and sunshine ravishing. Betty did not come along for altho it was not a Russian yet it was a rushing tour. I shall be back in Merion on March the 6th and teach the 8th and the 9th. Hope to see you then.

Kindest greetings to you and your master (?)

as always most sincerely yours,

Josef Hofmann

p.s. I shall try to arrange another recital or two for you but they should take place shortly before your appearance in N. Y. else they would do you no good. Please let me know "when"!

NR, William Harms, Joseph Levine, and Josef Hofmann

Hofmann would have been a bad teacher for anyone who was not a ready an advanced musician. He was no pedagogue. I had two hour lessons with Hofmann every week, and each time I brought new material because when—in the beginning—I brought the same piece twice, his comments would be completely the opposite of what he had said before. But that wasn't important because I was already a mature, educated musician, and could either accept or not accept his advice.

NR, 1981

What Hofmann taught me about pedalling, however, was priceless. I learned from him what I call college pedalling—using a combination of foot and finger pedalling, taking a pedal diminuendo, taking half, even quarter pedal in long phrases.

NR, 1981

Family Interlude II

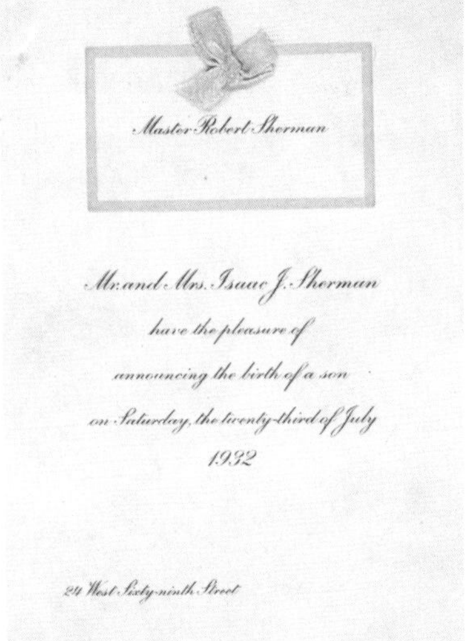

Master Robert Sherman

Mr. and Mrs. Isaac J. Sherman
have the pleasure of
announcing the birth of a son
on Saturday, the twenty-third of July
1932

24 West Sixty-ninth Street

NR with second son, Robert, born July 23, 1932

NR with second son, Robert

Wedding anniversary card from Alex to NR and Sasha, June 24, 1934

Meir Sherman (Sasha's nephew and Newta's husband), and NR

Nadia was attractive above all because of her tremendous vitality, her bursting,energetic, sometimes nervous lust for living, and the joy of living. But there was warmth in her, and kindness, and selfless devotion. She was more than a pianist, she was a musician, one who lived steeped in music, who could never tire of it—to make it or listen to it. She approached music with earnestness and humility.

Meir Sherman, 1938

Concert of Chamber Music of Johannes Brahms

SIMEON BELLISON
Clarinet

NADIA REISENBERG
Piano

FELIX SALMOND
Violoncello

TOWN HALL
113 West 43rd Street

MONDAY EVENING
NOVEMBER 6, 1933
at 9:00 o'clock

For the Benefit of the Committee on Ensemble Musical Training and Scholarships of the Philharmonic Symphony Society of New York.

Steinway Piano

CONCERT MANAGEMENT ARTHUR JUDSON, Inc.
Division of
Columbia Concerts Corporation of Columbia Broadcasting System
Steinway Hall 113 West 57th Street New York City
(over)

Chamber music is my first and real love in this world. Every time I play chamber music, it's a holiday. It's a marvelous, marvelous feeling; to play chamber music is to take yourself out of what *you* can do and into what *we* can do, and it's wonderful.

NR, 1977

[The Town Hall]

SEASON 1933-1934

Monday evening, November 6th, at 9 o'clock

•

CONCERT OF CHAMBER MUSIC

of

JOHANNES BRAHMS

Miss Nadia Reisenberg, *Piano*
Mr. Simeon Bellison, *Clarinet*
Mr. Felix Salmond, *Violoncello*

•

PROGRAM

I.

Sonata for Clarinet and Piano in F minor, Opus 120, No. 1
Allegro appassionato
Andante un poco Adagio
Allegretto grazioso
Vivace

Program Continued on Second Page Following

New York Times, 11/7/33

MUSIC

Brahms Chamber Music.

Whatever "benefit" may subsequently accrue to the ensemble training and scholarships organization of the Philharmonic-Symphony Society from last night's chamber music concert in the Town Hall, the first beneficiary was undoubtedly the audience. One seldom has the opportunity to hear Brahms's sonata for clarinet and piano and his trio for these instruments and 'cello. Nadia Reisenberg, pianist; Felix Salmond, 'cellist, and Simeon Bellison, clarinetist, opened and closed a charming concert with these works, and 'cellist and pianist gave an admirable reading of Brahms's sonata in E minor, op. 38.

The three musicians were at their best perhaps in the andantino of the trio. In his chamber music one comes close to the curiously diverse strains in Brahms's musical thinking; the warp of tenderness woven across the woof of tempestuous asperity. In the smaller forms, such as those set forth last night, these contrasts, this play of gentle light across architectural forms rugged and imposing, are very striking. The players projected them for the most part with a fine integrity, and in the andantino, with a faultless balance of tonal masses. Sometimes, although their grasp of outline and spirit was never at fault, this balance could have been improved, as in the opening allegro of the trio, and in parts of the 'cello sonata. Here, such is Brahms's fondness for thickened sonorities—the frequent presence of the third in lower registers—the 'cello sometimes suffered from a too prominent piano part, although the introductory "allegro non troppo" was the high point of the evening in poetic sensibility and sheer loveliness of tone and nuace.
loveliness of tone and nuance.

The audience, a good-sized one, applauded the players warmly.

H. H.

To Nadia Reisenberg, to the Artist, to the Colleague... In Friendship and Admiration of our long and successful ensemble playing.
S. Bellison, New York, 1939

PIANO RECITAL

NADIA REISENBERG

TOWN HALL

MONDAY EVENING at 8:30 o'clock

JANUARY 29, 1934

PROGRAM

I.

SONATA in D major .... SCARLATTI

SUITE in D minor, No. 2 .... BACH-GODOWSKI

Prelude—Allemande—Courante

Sarabande—Menuet I—Menuet II—Gigue

II.

SONATA in F-sharp minor .... SCHUMANN

Introduzione—Un poco Adagio—Allegro vivace

Aria

Scherzo ed Intermezzo—Allegrissimo

Finale—Allegro un poco maestoso

III.

RIGAUDON } (from "Tombeau du Couperin")

MENUET }

ALBORADA DEL GRACIOSO (from "Miroirs") .... RAVEL

ETUDE in F-sharp major .... STRAWINSKY

ETUDE in D minor .... PROKOFIEW

IV.

RHAPSODIE ESPAGNOLE .... LISZT

Tickets: Orchestra $2.20, $1.65; Balcony $1.10, $.83 (tax included) • Steinway Piano

CONCERT MANAGEMENT ARTHUR JUDSON, Inc., 113 West 57th Street, New York City

Division of Columbia Concerts Corporation of Columbia Broadcasting System

NADIA

REISENBERG

"A pianist of ripe musicianship."

L. Liebling in the N. Y. American.

Praised by New York press after her recital Jan. 29, 1934.

"Abundance of vitality — strong sense for color. Rhythmic suppleness. She gave a particularly charming reading of Ravel pieces, a reading beautifully molded and curiously touching."

N. Y. Times, Jan. 30, 1934.

"Crystal clarity of tone."

N. Y. Evening Post, Jan. 30, 1934.

"Talent of high order."

N. Y. Evening Journal, Jan. 30, 1934.

"Incisive strongly rhythmic playing."

N. Y. Sun, Jan. 30, 1934.

"Fine sense of rhythm—knowledge of style—grace and transparence. Ardent temperament under intelligent control."

N. Y. World Telegram, Jan. 30, 1934.

"Delicate, and her grasp of the content of music of the classic school that of a mature artist."

Brooklyn Daily Eagle, Jan. 30, 1934.

CONCERT MANAGEMENT
ARTHUR JUDSON, INC.
Steinway Building, New York
Division: Columbia Concerts Corporation of Columbia Broadcasting System

Steinway Piano

New York American. n.d.

NADIA REISENBERG WINS MUCH FAVOR AT PIANO RECITAL

By LEONARD LIEBLING

Long known here as a pianist of ripe musicianship and complete technical accomplishments, Nadia Reisenberg gave a recital at Town Hall last evening and won renewed response from a large audience of her admirers.

Miss Reisenberg's program in itself was a welcome departure from the worn ways of the usual pianistic repertoire heard at public concerts these days. She presented an excellently balanced list consisting of Mozart's "Pastorale Variee", sonata; the Suite, D minor, No. 2, by Bach-Godowsky; Schumann's F sharp minor sonata; Ravel's "Rigaudon", and "Menuet", "Alborado del Gracioso"; Stravinsky's "Etude", in F sharp; Prokofieff's "Etude", in D minor; and the "Spanish Rhapsody" by Liszt.

Articulated with transparency and done with delicate touch, the Mozart music served as an ingratiating prelude to the more serious business of the Bach pages so typically edited by that master transcriber, Leopold Godowsky. He was in the audience last evening and must have been pleased at the penetrative understanding and fine dignity with which his arrangement was treated

She did justice in the main also to the glorious but far too seldom played sonata by Schumann. The only lack in the performance was a tendency toward constructional exposition at the expense of some of the romantically passionate content. However, atonement was made by the vim and rhythmic life of the scherzo and the finale.

Many graces of tone and interpretative nuance were in the concluding brilliant group.

When I first came to the U.S., I premiered pieces by Shostakovich, Stravinsky, Miaskovsky and a great many other—at that time—modern composers; so I accept 20th-century music very easily, and use a great deal of Bartok, Stravinsky, and so on, in my teaching.

When it comes to ultra-modern music, it's a little different. Not that it shocks me, not that I'm disturbed by it, I just don't get the message, so it doesn't satisfy me.

It's very possible that if I live another ten or fifteen years I'll accept it very easily. With time, our tastes change, our ears get used to things. I still can't accept rock, though. What can I do? I'm not very up-to-date...

NR, 1981

NADIA REISENBERG

introduced a recent piano work by Dmitri Shostakovitch, composer of Lady Macbeth of Mtsensk, at the League of Composers concert on April 15 in the French Institute, New York.

New York Post, 4/20/35

New York Post, 11/12/33

Thanks From Schoenberg

ARNOLD SCHOENBERG made a little speech of thanks to those who applauded his music at Saturday night's concert in the Town Hall, given in his honor by the League of Composers.

Rita Sebastian's mood projection in four of the early songs of Opus 6, with accompaniments played by Edna Shepherd, and Nadia Reisenberg's sacrificial piano artistry in Opus 11 and Opus 33 was also to be admired.

Musical Courier, 11/18/33

NOVEMBER 11

League of Composers (Schönberg Concert) — On Saturday evening the League of Composers arranged a concert at Town Hall in honor of Arnold Schönberg, and after the finish of the program tendered a reception to him in the club rooms of the edifice.

The program proper, all Schönberg compositions, presented the string quartet No. 2, with voice; the quartet No. 3; and piano pieces and songs. The Pro Arte Quartet played the ensemble numbers. Rita Sebastian, contralto, and Ruth Rodgers, soprano, did the singing; Nadia Reisenberg contributed the works for solo piano; and Edna Sheppard accompanied the contralto.

At this late day no extended analysis of Schönberg's various styles and tendencies, is necessary for readers of the Musical Courier. They know of his early "Wagnerian" manner, and his later experiments in atonality and avowedly mathematical methods of composition. The audience that listened last Saturday was of a kind to follow the Schönberg tendencies with consideration and understanding. He was encouraged by the applause to come upon the stage and give thanks with a modest and hesitant short speech delivered in painstaking English.

The third string quartet, splendidly performed by the Pro Arte players, rewarded the concentrated attention of serious students of the Schönberg muse. The work has form and purpose, and its content delivered a recognizable message, even if it still sounds remote to those who prefer accepted harmonic euphony and obvious tonality, to unconventional and frequently acrid combinations of chords and sequences. Of the second string quartet much the same may be said. Its two last movements employ the human voice in addition to the string instruments, but the mournful songs (poems, called Litanei and Entrückung, by Stefan George) are abstruse, and seem the result of mental impulse rather than emotional stimulus. Miss Rodgers' singing was intelligent and musical. A satirical moment in the quartet—its difficulties again showed the Pro Arte players in a distinguished light—was Schönberg's sportive quotation of the tune of Ach. du lieber Augustin, which brought the only merry moment of the concert.

The four songs (Traumleben, Verlassen, Chasel, and Der Wanderer) date from Schönberg's earlier career, and have in them something of the influences of the classic Lied writers and of Wagner, although here and there are acidulous harmonies and a jumpy melodic line which indicate the dawning of the later revolutionary in music. Miss Sebastian's interpretative instinct, pure intonation, and clear utterance of text, were as welcome as the smooth flow and rich quality of her lovely voice.

Nadia Reisenberg lent to the three piano pieces, op. 11, and the Klavierstücke, op. 33, all the resources of her fluent technic and experienced musicianship, but to the majority of hearers those works continue to be unsensuous, strange, and disconcerting in design and result.

The celebrated Schönberg has joined the faculty of the Malkin Conservatory of Music, in Boston, and taken charge of its department of theory and composition.

Konserthuset, Lilla Salen, fredagen den 26 februari 1937

NADIA REISENBERG

MOZART . . . . Pastorale Variée.

SCHUMANN . . . . Sonat, fiss-moll
Introduzione — Un poco adagio — Allegro vivace.
Aria.
Scherzo ed Intermezzo — Allegrissimo.
Finale — Allegro un poco maestoso.

CHOPIN . . . . Ballad, Ass-dur
Nocturne, op. 55, n:r 2
Vals, Ass-dur

— PAUS —

RAVEL . . . . Rigaudon, Menuet — ur »Tombeau de Couperin»
Alborada del Gracioso (ur »Miroirs»)

STRAVINSKY . . . . Etyd, Fiss-dur

PROKOFIEFF . . . . Etyd, d-moll

LISZT . . . . Etyd, f-moll (ur »Douze études d'exécution transcendante»)

Steinwayflygel från Lundholms Pianomagasin

HÆGGSTRÖMS, STHLM 371030

Se sista sidan!

At the Franz Liszt Academy; Hermann Scherchen, conductor

CLARA ROCKMORE

CLARA ROCKMORE

Program

Concerto in E minor . . . . NARDINI
Allegro moderato
Andante cantabile
Allegretto giocoso

Sonata in A major . . . . CESAR FRANCK
Allegretto ben moderato
Allegro
Recitative—Fantasia
Allegretto poco mosso

INTERMISSION

Hebrew Melody . . . . ACHRON
Valse sentimentale }
Berceuse } . . . . TSCHAIKOWSKY
Pièce en forme de Habanera . . . . RAVEL
March of the Night Guard . . . . KORNGOLD

STEINWAY PIANO

Tickets: $2.20, $1.65, $1.10. Loges, seating six, $16.50

PRESS EXCERPTS

NEW YORK WORLD-TELEGRAM
"Miss Rockmore as has been noted before is an excellent thereminist, but first of all she is an excellent artist." (League of Composers concert).

NEW YORK SUN
"There was also much to admire in Clara Rockmore's skillful performance on the space-controlled instrument. Her playing was consistently accurate in pitch and possessed of dynamic gradations completely individual to the instrument." (League of Composers concert).

SOLOIST WITH PHILADELPHIA ORCHESTRA AT ROBIN HOOD DELL

PHILADELPHIA PUBLIC LEDGER
"Miss Rockmore, originally a violinist and a pupil of Kneisel, Auer, and other great masters, proved to be a master of the theremin . . . she played the difficult "Schelomo" of Ernest Bloch and scored an immense success."—*Samuel L. Laciar*

PHILADELPHIA EVENING BULLETIN
"Miss Rockmore is the greatest virtuoso of the instrument . . . she has come to the work with extraordinary qualifications. With an assurance which must have been born of instinct as well as skill, she calmly darts her fingers at invisible spots in the atmosphere and produces the desired pitches with wonderful accuracy."

PHILADELPHIA INQUIRER
"Miss Rockmore had full command of her solo instrument and utilized every opportunity for exploitation of the emotional nature of 'Schelomo.' Bloch's rhapsody was originally written for 'cello but is excellently suited to the brilliance of the theremin."

SOLOIST WITH TORONTO SYMPHONY ORCHESTRA

THE EVENING TELEGRAM, TORONTO
"There was enchantment in the playing of Clara Rockmore . . . tenseness of mood and vivid imagination . . . it was most beautiful."

THE MAIL AND EMPIRE, TORONTO
"The tone was extremely rich resembling that of the 'cello, but with much greater volume and a very romantic timbre . . . it combined very effectively with orchestra . . . the acumen of the soloist in intonation and interpretation was most impressive."

Management: METROPOLITAN MUSICAL BUREAU, INC.
Division of Columbia Concerts Corporation of Columbia Broadcasting System
113 West 57th Street New York City

Nadia was a perfect colleague. There was hardly a need for rehearsals because she knew what I wanted to do before I did it. I could play with complete freedom and she was always right there with me...

Clara, 1984

I had already played "Schelomo" with the Philadelphia Orchestra by then, so the theremin was not exactly an unknown instrument; still a full Town Hall recital was definitely a test of some kind. Nadia worked with me on the program every step of the way, and playing the complete Franck Sonata remains one of the high points of my life. We approached it with such love and pride, and even though officially it was my recital, not once would I take a bow without her. That may seem like a small thing, but it was very important to me...

Clara, 1985

New York Times, 2/27/45

MISS ROCKMORE GIVES RECITAL ON THEREMIN

Offers Skillful Performance on Ether-Wave Instrument

An effective demonstration of the present capabilities of the theremin ether-wave instrument in the hands of a fine musician who has fully conquered its difficult technique, was made by Clara Rockmore at her recital last night in Town Hall. What Miss Rockmore accomplished in matters of intonation, legato and staccato playing, accentuation and nuances, must have come as a distinct surprise to any in the house who entertained false ideas of the theremin's possibilities.

Doubtless the improved security in pitch was due to improvements brought about by its inventor during the last two years, due largely to suggestions made by Josef Hofmann, the celebrated pianist. One was a method of preventing the tubes from becoming warm too rapidly, which has resulted in less frequent need of tuning.

That the theremin can respond cleanly and nimbly in speedy passages when the performer has sufficient command of its resources became patent in Miss Rockmore's agile treatment of the second movement of the César Franck violin sonata, which was taken at as high a rate of speed as violinists ever adopt for its allegro sections. In this formidable work and the preceding Nardini concerto in D minor, Miss Rockmore evinced not only her exceptional control of every phase of theremin playing, but also a highly developed musical imagination and estimable interpretative powers. She was deftly assisted by Nadia Reisenberg at the piano. The audience was warm in its approval. N. S.

New York Post, 2/27/45

CLARA ROCKMORE THEREMIN RECITAL

Last night in Town Hall a large audience sat enthralled while Clara Rockmore, assisted by Nadia Reisenberg at the piano, played a recital on the electrical instrument, the Theremin. The program included a Concerto in E minor by Nardini; the Franck Sonata in A major, and a group of pieces by Achron, Tchaikovsky, Ravel and Korngold.

Standing before the cabinet of the instrument, with the speaker behind her, Miss Rockmore, in a manner suggestive of the supernatural, drew exquisite tones seemingly out of thin air. With no tangible point of contact with the Theremin she performed difficult scale passages and leaps with an astounding dexterity and precision. The instrument, which is not comparable to any other existing one, has a great range of pitch and dynamics and appears to possess no technical limits. However, even if the woods were full of Theremin players, Miss Rockmore would still be outstanding, since she is a first-rate musician and a fine artist O'G.

Clara is a consummate musician, with impeccable taste, so we can attain wonderful ensemble almost without rehearsing...

NR, 1978

TOWN HALL

Sunday Afternoon at 3:00 o'clock

NOVEMBER 21, 1937

PROGRAM

I.

Tempo di Ballo }
Toccata } .... SCARLATTI

Novellette, Opus 21, No. 8 .... SCHUMANN

II.

Sonata in D major, Opus 53 .... SCHUBERT

Allegro vivace
Con moto
Scherzo—Allegro Vivace
Rondo—Allegro Moderato

INTERMISSION

III.

Intermezzo in B-flat minor, Opus 117 }
Intermezzo in A major, Opus 118 }
Rhapsodie in E-flat major, Opus 119 } .... BRAHMS

IV.

Suite .... FRANCIS POULENC

Presto—Andante—Vif

Three Etudes, Opus 8 .... SCRIABINE

F-sharp minor
E minor
D-sharp minor

Etude Transcendante in F minor .... LISZT

Steinway Piano

Tickets: Orchestra A-K $2.20, L-U $1.65; Balcony A-F $1.10, G-M 83c • Loges, seating six, $16.50

CONCERT MANAGEMENT ARTHUR JUDSON, INC.
Division of Columbia Concerts Corporation of Columbia Broadcasting System
Steinway Hall 113 West 57th Street New York

New York Times, 11/22/37

NADIA REISENBERG HEARD IN RECITAL

Russian Pianist's Program at Town Hall Is Applauded by Large Audience

EXACTING MUSIC PLAYED

Schubert Sonata Is the Chief Number—Brahms Group Also Is Included

Nadia Reisenberg, Russian pianist, gave one of her infrequent recitals yesterday afternoon at Town Hall. Miss Reisenberg came to this country as a refugee after the World War while still in her 'teens and was first heard in New York in the early twenties. Since then she has appeared here often both as soloist and in various types of ensemble work. Her performances yesterday were enthusiastically received by a large audience, including numerous prominent musicians.

Miss Reisenberg's playing, as always, possessed brilliance and sparkle. Her fluent, facile finger-technique made it possible for her to skim over the keys accurately and glibly in the most perilous pages on her list without sacrificing clearness and precision. The speedy tempi adopted for the Scarlatti sonatas converted them into feats of bravura. Nor was virtuosity lacking in the Schumann "Novellette," or the Schubert sonata.

Extremes of tone were favored by the pianist, delicately and pleasingly manipulated soft passages giving way to full-blooded fortissimo effects, where the tone was produced with effort and therefore was not invariably of the best quality. In the Schubert sonata this kind of coloring proved insufficient in an interpretation which rarely went far beneath the surface emotionally. The rubato introduced, if indulged in for expressive purposes, did not quite serve its purpose, nor could one see eye to eye with Miss Reisenberg when it came to the matter of shaping phrases, hers having a peculiar manner of slumping at climactic points. But in all she accomplished, the artist had very definitely defined ideas of what she wanted to do and the ability to carry them vividly across the footlights.

N. S.

List of Mozart Piano Concertos broadcasted by
NADIA REISENBERG

Date		Concerto	K.
Sept. 12,	1939	No. 1 in F	K.37
" 19,	"	No. 2 in B flat	K.39
" 26,	"	No. 3 in D	K.40
Oct. 3,	"	No. 4 in G	K.41
" 3,	"	and No. 1	K.107

K.107 broadcast on October 3 and 10 comprises the three Sonatas by Joh. Crh. Back arranged as piano concertos by Mozart.

Date		Concerto	K.
" 10,	"	Nos. 2 and 3	K.107
" 17,	"	No. 5 in D	K.175
" 24,	"	No. 6 in B flat	K.238
" 31,	"	No. 7 in F for 3 Pianos	K.242
Nov. 7,	"	No. 8 in C	K.246
" 14,	"	No. 9 in E flat	K.271
" 21,	"	No. 10 in E flat for 2 Pianos	K.365
" 28,	"	Concert Rondo in D	K.382
		in A	K.386
" 28,	"	Concert Rondo	
Dec. 5,	1939	No. 11 in F	K.413
" 12,	"	No. 12 in A	K.414
" 19,	"	No. 13 in C	K.415
" 26,	"	No. 14 in E flat	K.449
Jan. 2,	1940	No. 15 in B flat	K.450
" 9,	"	No. 16 in D	K.451
" 16,	"	No. 17 in G	K.453
" 23,	"	No. 18 in B flat	K.456
" 30,	"	No. 19 in F	K.459
Feb. 6,	"	No. 20 in D minor	K.466
" 13,	"	No. 21 in C	K.467
" 20,	"	No. 22 in E flat	K.482
" 27,	"	No. 23 in A	K.488
Mar. 5,	"	No. 24 in C minor	K.491
" 12,	"	No. 25 in C	K.503
" 19,	"	No. 26 in D	K.537
" 26,	"	No. 27 in B flat	K.595

"K" refers to the number of the respective works as listed in Kochel's Chronological Thematic Catalog of Mozart's Musical Compositions.

Musical History

is being made by the concert pianist of international reputation

Nadia Reisenberg

accompanied by the WOR Symphony Orchestra under the baton of the Musical Director of Station WOR

Alfred Wallenstein

in a Series of twenty-nine weekly broadcasts of

All of Mozart Piano Concertos

in the chronological order of their composition.

NEVER BEFORE has a task of such magnitude been attempted in musical history.

It is indeed a great contribution to the musical life of America and a veritable *tour-de-force* of the brilliant pianist

Nadia Reisenberg

BROADCASTING

every TUESDAY EVENING

9:30 to 10:00 Eastern Standard Time over the Station **WOR** or the Mutual Broadcasting System.

It was the most rewarding experience of my career, my private year with Mozart! Sometimes I would be besieged by all the tunes—I always had six concertos on my piano at a time: finishing two, working on another two, and reading through a third pair—and all of that month after month. At times I was in such a state of agitation that I couldn't work or sleep, so then I would take the car in the evening and drive into the country for an hour or two. Poor Dad had to come along and suffer, but the steady motion of the car relaxed me, and I would return home rested and cheerful.

NR, 1974

Of all the composers, I find Mozart the most satisfying—perhaps I shall make the exception of Bach. Mozart does not allow mediocrity, not even one false note: his kind of divine inspiration must be crystal clear. The more I played the concertos, the more I became attached to them; I put each one away with real regret and a promise to myself that I would return to it soon—I have a tenderness for them all.

NR, 1940 (*Keyboard* Magazine)

The Pianist's Mozart

NADIA REISENBERG

Concert Pianist par excellence; soloist for WOR's Mozart Festival

Where lies Mozart's genius? How approach him at the piano? One of our outstanding Mozart interpreters gives a richly stimulating analysis.

THE amazing thing about Mozart's compositions for the piano is the fact that they appear to be easily readable to any fairly advanced student. . . . And yet, I dare say, in many respects it is more difficult to play the "easy" Mozart than the most complicated modern compositions. How much easier it is to thunder away on the piano some bombastic piece! And how inconceivably hard it is to bring out the simplicity and serenity of a Mozartian phrase!

What, precisely, are the difficulties in playing Mozart, from a purely mechanical point of view?

MOZART, more than any other composer, requires *perfect poise*, which cannot be achieved without complete control of *all* the fingers of each hand.

The fourth and fifth fingers are usually insufficiently developed, because of their unfavorable position. For the same reason, the thumb is often clumsy and heavy. . . . As a result, a passage will lose its smoothness and evenness; one note may sound louder than another, when all notes are meant to sound alike. It is unbelievable how the slightest imperfection will distort a thought and put a phrase out of shape.

The weakness of the fourth and fifth fingers and the clumsiness of the thumb are, of course, basic defects which would adversely affect the playing of any other composer. However, they are particularly noticeable in the playing of Mozart, because of the extreme transparency and lucidity of his music.

FINDING the proper *speed* with which to play rapid passages is quite a problem. . . . There is a temptation to play them as fast as the technical equipment permits, thus treating these passages as if they are brilliant runs of a purely ornamental nature. But what is appropriate and laudable in playing a Light Rhapsody is out of place when playing Mozart.

Practically all his passages—even in the fast movements—are of melodic nature and, in most cases, convey a *thought* rather than a *mood*—a message which has to be carried and made clear to the listeners. And just as all fingers must move absolutely evenly, so the dynamics must be planned in such a manner that all crescendos and diminuendos be perfectly timed.

When playing fast passages in some modern compositions—such as Prokofieff—it is sometimes more important to convey a mood than a thought. To achieve a certain mood, one may have to plan the dynamics in such a way that the crescendos and diminuendos come with a certain abruptness—the opposite of what the structural design calls for—and several notes of a given passage have to be blurred on purpose, like the colors in an impressionistic painting.

In most of Mozart's passages, every note must be given its full value. One must take as much care of the last sixteenth of a given quarter as of the first, so as to ensure a perfect completion of a phrase or a line.

THE finger pressure on the keys must be properly distributed, so as to achieve a perfect articulation, which is essential for Mozart. An ear capable to detect the slightest unevenness and a muscular coordination ready at a flash to adjust it—these are of utmost importance.

A great deal of thought must be given to skillful and ingenious *fingering*. I am not partial to any fingering given in a particular edition. . . . I believe that fingering must be strictly individual and depend on the strength of the fingers, size of the hands, and other peculiarities of each person.

Some players have a tendency to take certain passages too slowly—making them dragging and sentimental. Others take

(Continued on page 46)

The Pianist's Mozart
(Continued from page 7)

passages too fast—robbing them of their significance and distorting them into ornamental runs. Each passage, of course, has its own right tempo, but—generally speaking—I may say that one has to play each passage in a tempo which will allow *all the notes to be played out, all phrases to be completed, all the musical contents expressed,* without hurrying, with good balance, with poise and complete self-control.

PEDAL must be used sparingly and with utmost caution in the interest of precise articulation and clarity of sound.

One word about Mozart *sonority*. It is of transparent lightness and airiness, and differs radically from the sonority of, let us say, Beethoven, which is more robust and meaty. In the treatment of Cantilena, the difficulty consists in achieving warmth and depth of tone, yet keeping it within the classical style, without the sweep and subjectivity of the romantic school.

AS to the Mozart style, it is impossible to define it adequately in words. This is a matter of feelings, taste and musical intelligence.

If I should try to express my feeling about it, I would emphasize as essential features a few points which appear to me characteristic.

Mozart's style has an elegance and a certain polished refinement. This applies even to his more serious and often tragic moods. . . . There is also purity of thought, spiritedness and vitality.

Mozart is joyful and happy. He may be extremely sad at times, he may be tortured by unhappy thoughts, but at the end he affirms life, regains the peace of soul. . . . He may be in despair but he is never morbid.

All Mozart compositions have a deep, intense feeling. He is never indifferent, apathetic, frivolous or cold. He feels deeply, warmly, intensely.

And then there is the atmosphere of an intimacy which seems to permeate all Mozart's compositions.

SOME play Mozart somewhat effeminately. . . . I think this is wrong. Mozart is tender; he is sensitive, he has a great deal of sentiment. But he is not weak; he is never sentimental.

Some try to make him sound robust and over-vigorous, but Mozart is distorted by such over-emphasis. He is spirited, he has tremendous vitality—but he is neither coarse nor brutal.

. . . However, the style of a composer cannot be satisfactorily explained by mere words. I would very much prefer to present to my readers my conception of and my feeling for Mozart style by performing for them some of the works of this magnificent genius.

Some of my happiest moments were the broadcasts with Nadia Reisenberg. That was an era we will probably never see again: all the broadcasts were live, without commercials, and splendid musicians were playing great music for millions of people. Nadia was the first pianist in our country to have publicly performed all of the Mozart concertos in sequence, and it was quite a feat.

Alfred Wallenstein, 1974

The weekly schedule did not leave much time for thinking and analyzing, so I had to depend on my intuition and feeling for Mozart; this probably made my performances more spontaneous, more emotional and sincere than they might have been under different circumstances.

NR, 1940

Never before did I feel so close to the innermost substance of music. The purity of form, the loftiness and melodic wealth of Mozart were enthralling. I can truly say that mine was a work of love and enthusiasm, and the series was a source of infinite joy. I shall always cherish those seven months as the happiest time of my entire musical life.

NR, 1940

She played brilliantly and I admired her very greatly. I'll never forget the cycle of concertos which revealed to me lots of Mozart beauty that I keep in my heart with much gratefulness to her.

Artur Rubinstein, 1979

For Nadia,
a great artist, and wonderful woman—
Sincerely, Alfred Wallenstein 10/14/41

The 1940s

New Friendships, New Horizons

Sir John Barbirolli

C/O The Hotel Devonshire
Vancouver
B.C.

29/5/40

My dear Nadia Reisenberg,

Re your appearance with the Philharmonic, would you please consider very sympathetically the following suggestion:
I am very anxious to introduce to the public an excellent new piano concerto by Mischa Portnoff, and I would like to ask you to play it. I know how you will feel about making your appearance with a brand new work, so if you would do the Portnoff(he would let you have it at once to look at, or could play it to you), I could arrange for you to make a second appearance in the same programme with perhaps the Fantaisie of Debussy of which we spoke so enthusiastically together some while ago.

I do hope that you are at last enjoying a very well earned holiday. My wife and I have taken a house near here right up in the hills. It is wonderfully beautiful country. and very varied, since we have sea woods, country lanes and mountains all just nearby!
We both send you all our good wishes for a most pleasant and restful summer,

Very sincerely yours,

John Barbirolli

New York Times, 2/25/41

PREMIERE IS GIVEN BY PHILHARMONIC

First Performance of Mischa Portnoff's Piano Concerto Heard at Carnegie Hall

NADIA REISENBERG PLAYS

Soloist in New Work — Liszt Composition and 'Freischutz' Overture Are Offered

By OLIN DOWNES

John Barbirolli, in the course of one of the best concerts he has conducted here, led the New York Philharmonic-Symphony Orchestra yesterday afternoon in Carnegie Hall in the first performance of Mischa Portnoff's Piano Concerto. The solo part was played by Nadia Reisenberg. The performance was received with prolonged applause.

Miss Reisenberg must have gratified the composer, who appeared on the platform with her after the performance, by the virtuosity and command that she showed in her mastery of a piano part that called not merely for physical deeds of derring-do but also for musical thinking of a highly concentrated kind. The complexities of her task did not blind her at any time to the essential proportions of the music.

1940 *Ninety-ninth Season* 1941

THE PHILHARMONIC-SYMPHONY SOCIETY OF NEW YORK

at CARNEGIE HALL

Thursday Evening, February 20, at 8:45
Friday Afternoon, February 21, at 2:30

Premiere of New Version of Opera
"CYRANO DE BERGERAC"

Conducted by the Composer
WALTER DAMROSCH
Book by W. J. HENDERSON
After the Drama by EDMOND ROSTAND

Under the Direction of
JOHN BARBIROLLI

Sunday Afternoon, February 23, at 3:00
Assisting Artist:
NADIA REISENBERG, *Pianist*

WEBER.......Overture to "Der Freischütz"
PORTNOFF.......Concerto for Piano and Orchestra
(*First Performance Anywhere*)
DELIUS.......Prelude and Serenade from "Hassan"
LISZT.......Piano Concerto in A major, No. 2
DUKAS.......Scherzo, "The Sorcerer's Apprentice"

ARTHUR JUDSON, *Manager*
BRUNO ZIRATO, *Associate Manager*

The STEINWAY *is the Official Piano of The Philharmonic-Symphony Society*

CARNEGIE HALL Friday Evening, February 21st, at 8:30 o'clock

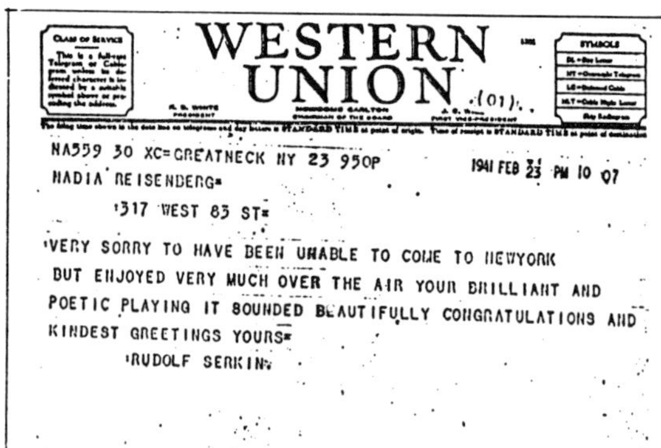

WESTERN UNION

NA559 30 XC=GREATNECK NY 23 950P 1941 FEB 23 PM 10 07

NADIA REISENBERG=
317 WEST 83 ST=

VERY SORRY TO HAVE BEEN UNABLE TO COME TO NEWYORK BUT ENJOYED VERY MUCH OVER THE AIR YOUR BRILLIANT AND POETIC PLAYING IT SOUNDED BEAUTIFULLY CONGRATULATIONS AND KINDEST GREETINGS YOURS=
RUDOLF SERKIN=

Soloists dream of a conductor who is sympathetic in listening, very sensitive in guessing their wishes. A conductor can be brilliant for purely orchestral work but—I don't want to say stubborn!—not sufficiently sensitive to the needs of the soloist. I in no way mean he should merely follow, of course—he has to show his own personality as well.

NR, 1978

A real conductor is a thorough musician, one who knows the score inside out, who understands the sonorities and balances of the orchestra, who knows how to rehearse. I didn't consider Barbirolli the greatest of conductors, but he was positively the greatest accompanist: he felt ahead of time what the soloist wanted—to take a little more time, to move ahead—and he went with them. He had a way of listening to you that let you be completely free.

NR, 1981

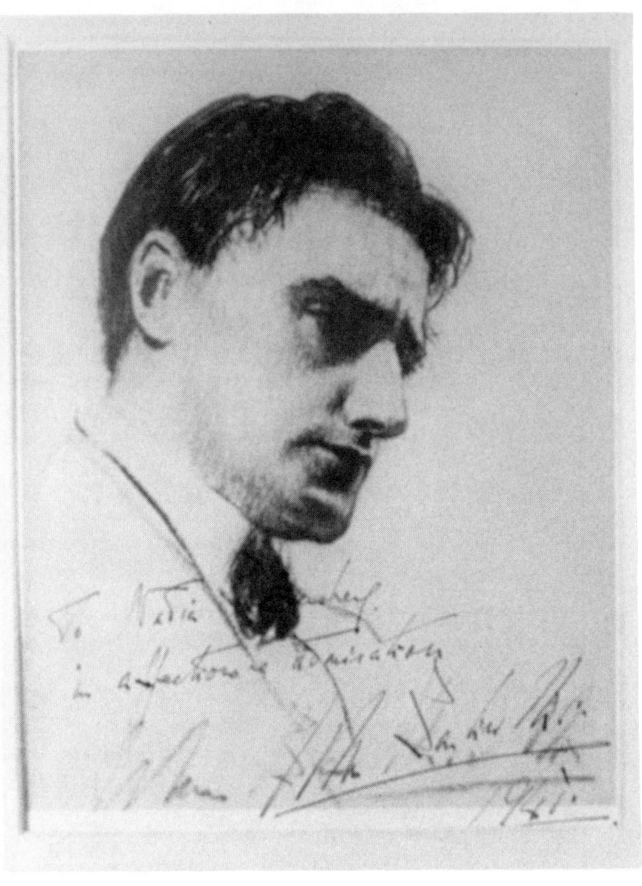

In these trying days through which England is passing, I can't help thinking of you and Mrs. Barbirolli. The news from London is heartrending and, if it is so for us, I can imagine how more intensely and deeply you must feel about it. However, true artist that you are, I somehow feel that you have succeeded in plunging yourself into your scores deeper than ever. Where can we escape if not into our art? Where can we find solace amid the horrors of war, but in music?

NR, letter to Sir John Barbirolli, 9-10-40

On the train
September 19, 1940

My dear Nadia Reisenberg:

We were both very deeply touched by your kind thought in writing us. It is truly an agonizing time for us to be passing through with all our loved-ones in the midst of this desperate battle. So far at least, up to a few days ago, they were still all right, and even since the beginning of these horrible raids have come messages of cheerful courage and determination. We cannot help feeling great pride in the achievement of our people, and somehow in our hearts we know such courage and determination and the cause of all that is decent and human must one day have its reward.

I am glad to know that you look forward to the D'Indy. It is indeed a very lovely work and requires the collaboration of a fine musician as well as pianist.

I was glad to hear that you are having something of a holiday in the midst of all your work, and please tell Schuster I hope his cello practise is not interferring with his ping-pong technique.

I have had a wonderful summer as far as concerts go,, and as usual, work has proved to be the best solace for all troubles.

With warmest greetings to you and your husband and to any friends that may be with you, believe me

Yours sincerely,

THE PHILHARMONIC-SYMPHONY SOCIETY OF NEW YORK

1842 1878

CONSOLIDATED 1928

• *1940 Ninety-Ninth Season 1941* •

CARNEGIE HALL

*Wednesday Afternoon, April 9, 1941
AT TWO-THIRTY

Thursday Evening, April 10, 1941
AT EIGHT FORTY-FIVE

3749th and 3750th Concerts

Under the Direction of

JOHN BARBIROLLI

Assisting Artist:
NADIA REISENBERG, *Pianist*

PROGRAM

HAYDN — Symphony in F minor, No. 49 ("La Passione")

I. Adagio
II. Allegro di molto
III. Menuetto
IV. Finale: Presto

WAGNER — Prelude to "Parsifal"

INTERMISSION

d'INDY — Symphony for Orchestra and Piano on a French Mountain Song, Op. 25

I. Assez lent, modérément animé, un peu plus vite—
II. Assez modéré, mais sans lenteur
III. Animé

NADIA REISENBERG

*Instead of Friday Afternoon

ARTHUR JUDSON, Manager — BRUNO ZIRATO, Associate Manager

MME. REISENBERG uses the STEINWAY PIANO

THE STEINWAY is the Official Piano of The Philharmonic-Symphony Society

COLUMBIA AND VICTOR RECORDS

ORCHESTRA PENSION FUND—*It is requested that subscribers who are unable to use their tickets, kindly return them to the Philharmonic-Symphony Offices 113 West 57th Street, or to the Box Office, Carnegie Hall, to be sold for the benefit of the Orchestra Pension Fund. All tickets received will be acknowledged.*

Nadia Reisenberg

The *Only Pianist Engaged Twice* in the *Same Season as Soloist* with the New York Philharmonic - Symphony Orchestra under John Barbirolli at Carnegie Hall

• **FEBRUARY 23, 1941**

Soloist in 2 Works:

Liszt A major Piano Concerto and Premiere of Piano Concerto by Mischa Portnoff

• *Olin Downes, New York Times*

"Miss Reisenberg must have gratified the composer by the virtuosity and command that she showed in her mastery of a piano part that called not merely for physical deeds of derring do, but also for musical thinking of a highly concentrated kind. The complexities of her task did not blind her at any time to the essential proportions of the music."

• *Jerome D. Bohm, New York Herald Tribune*

"Miss Reisenberg revealed extraordinary strength in her combat with the seething tonal masses behind her. She conquered the work's many technical difficulties and intricacies in masterly fashion and won the warm applause of the audience . . . Gave further proof of her virtuosity in the Liszt Concerto."

• *Oscar Thompson, New York Sun*

"There need be nothing but praise for the pianist."

• *Pitts Sanborn, New York World-Telegram*

"Its extremely exacting demands were met with signal ability . . . insistent applause followed . . . Later Mme. Reisenberg devoted her rare talents to summoning the magic of Liszt's A major Concerto."

• *Grena Bennett, New York Journal and American*

"Remarkably brilliant technical equipment and speedy, tireless fingers."

• **APRIL 9 and 10, 1941**

Soloist in

D'Indy Symphony based on a French Mountain Song

• *Louis Biancolli, New York World-Telegram*

"Virtuosity, beginning with Mme. Reisenberg, rode high in the brilliant D'Indy score. She upheld her prestige in sustaining an integral part in D'Indy's bustling scheme."

• *Irving Kolodin, New York Sun*

"Miss Reisenberg's playing of the important piano part was a feat of warm, sincere musicianship, properly scaled to the demands of the score."

• *Grena Bennett, New York Journal and American*

"Nadia Reisenberg, pianist, was the featured musician in D'Indy's symphony. She has earned her laurels as an admirably equipped musician and her reading of the solo part disclosed a definite and inherent sense of style and polished technique and an artful incorporation of accent. In each of the three movements she met the various exactions of the work with the assurance and confidence of a prime musician."

Re-engaged with the

NEW YORK PHILHARMONIC-SYMPHONY CENTENNIAL 1941-42

As Assisting Artist

in Bloch Concerto Grosso for Strings and Piano Obligatto and Scriabin's "Prometheus"

HAMPSHIRE HOUSE
150 CENTRAL PARK SOUTH
CIRCLE 6-7700

To Nadia,
To remember our "Symphonie Montagnard" & a great cause.
With affection & admiration.
John Barbirolli

To Nadia,
To remember our "Symphonie Montagnard" and a great cause.
With affection & admiration.
John Barbirolli

Joseph Schuster

Theirs was a deep, loving friendship, and their music-making would melt your heart...

Newta, 1985

YMHA PRESENTS

JOSEPH SCHUSTER, Cellist
NADIA REISENBERG, Pianist

IN TWO BEETHOVEN PROGRAMS
In All the Sonatas and Variations

WEDNESDAY EVENINGS, AT 9 P.M.
MARCH 4 and MARCH 25, 1942

THERESA L. KAUFMANN AUDITORIUM, 92nd St. and Lexington Ave., N. Y. C.
All seats reserved: $1.10

Exclusive Managements:
BOOSEY & HAWKES ARTISTS BUREAU, Inc., NEW YORK, N. Y.

THE MOUNT SINAI HOSPITAL

FIFTH AVENUE AND ONE HUNDREDTH STREET

NEW YORK March 10, 1942

Miss Nadia Reisenberg
317 West 83 Street
New York City

Dear Nadia:

I want to express to you again the thanks of the nurses of the Hospital and of myself for your share in making last night's concert such a very enjoyable event. And while we shall have a permanent souvenir of the occasion in the photographs taken, I should like to have one of your professional photographs with a suitable inscription that the nurses can hang in the room where they keep their greatest treasures.

March the ninth, 1942, may be just another concert date on your calendar, but to the nurses it will mean one of those rare evenings that they can write home about and, some day, tell their husbands and children about.

Sincerely yours

JOSEPH TURNER, M. D.
Director

S

168 EAST 71ST STREET

November 9, 1942.

Dear Nadia:

What a lovely combination of two fine artists! Your and Mr. Schuster's invitation is most kind, but this winter, for personal reasons, I have had to shut down on all concerts except two or three of the Philharmonic which I had promised to attend.

With cordial greetings from your eighty-year old but devoted friend,

Walter Damrosch

THE TOWN HALL

123 WEST 43rd STREET, NEW YORK, N. Y.

ALFRED SCOTT · PUBLISHER · 156 FIFTH AVENUE, NEW YORK

65-12-2E-42

New York Herald Tribune, n.d.

Beethoven's Works for Piano And Cello Given

Schuster-Reisenberg Team Appears in First of Series of 2 Town Hall Concerts

By Herbert F. Peyser

Guest Critic

Joseph Schuster and Nadia Reisenberg gave the first of a pair of concerts to embrace all of Beethoven's compositions for 'cello and piano at the Town Hall last evening. About a year ago they performed the identical series at a Y. M. H. A. and the report of that enterprise must have been flattering for an audience of uncommon size was on hand to welcome the artists yesterday. The actual undertaking is not as formidable as it may sound, for Beethoven's works for this combination are relatively few—five sonatas (as against twice that number for violin and piano) and three sets of variations. Yet it is good to hear these pieces, several of which are heard by no means as often as they deserve to be.

The sonatas in F major and G minor, Op. 5, written when the composer was twenty-six years old, certainly merit more attention than they ordinarily get. The program last night offered the first of the two.

The C major sonata, the first of the more or less enigmatic Op. 102 set, is a wholly different proposition and of much later date. Not even the polished and intelligent interpretation the two players' brought to it could make it seem, for all its weightiness, less hard-shelled or more musically alluring than it habitually appears. Nothing in it equals the profound and beautiful slow movement of its companion piece in D major.

Of the variations the two artists chose last evening, the twelve which Beethoven made on Papageno's "Ein Maedchen oder Weibchen" from the "Magic Flute." These are charming music, of early Beethoven vintage, not as deep, to be sure, as the great variations of a later period, but exhibiting unmistakably some of the typical features of Beethoven's variation technic. Disregarding chronology, Mr. Schuster and Mme. Reisenberg reserved the best for the close and terminated the concert with the great A-major sonata, not only the greatest by far of all the composer's creations for 'cello and piano, but one of his amplest and richest contributions to sonata literature from any standpoint. It was undoubtedly a fine sense of climax which induced the players to end the program with this masterpiece, but hardly kind to those reviewers whose duties drove them from the hall without an opportunity of hearing what promised to be a wholy distinguished performance.

Joseph Schuster, Shura Cherkassky, and NR

NR, Cherkassky, Schuster

Joseph Schuster, NR, and Mishel Piastro

NR, Piastro, and Schuster

SOCIETY OF ARRIVALS FROM EUROPE
and
EMERGENCY COMMITTEE FOR RELIEF

BARBIZON PLAZA
CONCERT HALL
101 West 58th Street

SATURDAY EVE
MARCH
20
1948
at 8:30 o'clock

Nadia REISENBERG	Mishel PIASTRO	Joseph SCHUSTER
piano	violin	'cello

Mishel PIASTRO	Nadia REISENBERG	Joseph SCHUSTER
violin	piano	'cello
Musical Director of the famous Symphonette "Artist of fine taste and distinction of style." *Laurence Gilman*	"One of the outstanding women pianists of the day." *N. Y. Times*	"Joseph Schuster is one of the most popular cellists of our time." *N. Y. Times, Jan. 1948*

PROGRAM

I

TRIO OP. 32 . . . . *ARENSKY*

Allegro moderato
Scherzo Allegro molto
Elegia Adagio
Finale — Allegro non troppo

Mishel Piastro — Nadia Reisenberg — Joseph Schuster

II

SONATA OP. 19 FOR CELLO AND PIANO . . . . *RACHMANINOFF*

Lento — Allegro moderato
Allegro Scherzando
Andante
Allegro mosso

Nadia Reisenberg — Joseph Schuster

Intermission

III

TRIO OP. 50 . . . . *TSCHAIKOWSKY*

Pezzo elegiaco
A. Tema con Variazioni
B. Variazione Finale e Coda

Mishel Piastro — Nadia Reisenberg — Joseph Schuster

Steinway Piano Used

Chamber music brings me closest to the pure sense of music-making. In playing chamber music one becomes more humble and less important than the work being performed. There is something so beautiful about making a phrase where the sense of teamwork, of give-and-take plays such an important role.

NR, 1974

Budapest String Quartet

When I was a young man, Nadia played with the Budapest Quartet all the time, and I used to listen to those performances *par excellence.* She is one of the great chamber musicians.

Robert Mann (Juilliard Quartet), 1978

When I played with the Budapest Quartet, what a pleasure it was to feel how beautifully we blended together, how wonderfully the phrase came out when each player gladly gave way to another because it's the other person's time to shine. That is something very pure, and very valuable to me...

NR, 1974

You don't just listen to your own playing, you listen to everyone. For example, in the second movement (of the Schumann Quintet) where the theme is played by every one of the musicians, I watch the bow of this one, the breathing of that one. It's a silent agreement...

NR, 1978

Nadia knows the Schumann Quintet in her sleep...and elsewhere!

Robert Mann, 1978

... *Gala Concert* ...

BUDAPEST STRING QUARTET

AND

NADIA REISENBERG

PIANO

Wednesday Evening, NOVEMBER 8, 1944 at 8:30 o'clock

Hunter College Assembly Hall

695 Park Avenue, New York City

... *Program* ...

I.

Trio in D minor.......................................Mendelssohn

Molto Allegro agitato
Andante con moto tranquillo
Scherzo
Finale—Allegro assai appasionato

Josef Roismann, Mischa Schneider, Nadia Reisenberg

II.

Barcarolle op. 60.......................................Chopin

Rhapsodie Espagnole.......................................Liszt

Nadia Reisenberg

—INTERMISSION—

III.

Quintet, Op. 44.......................................Schumann

Allegro brillante
In modo d'una Marcia. Un poco largamente
Scherzo—Molto vivace
Allegro, ma non troppo

Budapest String Quartet

Josef Roismann, 1st Violin *Edgar Ortenberg, 2nd Violin*
Boris Kroyt, Viola *Mischa Schneider, Cello*
and
Nadia Reisenberg, Piano

Columbia Records Steinway Piano Victor Records.

For the benefit of ORT TRADE SCHOOL

Tickets: \$3.60, \$2.40, \$1.80, \$1.20, 90c
Mail orders to ORT, 212 Fifth Avenue, New York 10, N.Y.

Exclusive Management for the Budapest String Quartet:
ANNIE FRIEDBERG, 251 W. 57 St., New York 19

In concert with Budapest String Quartet. Son Bob (left) is page turner

Artur Rodzinski

The Cleveland Orchestra
ARTUR RODZINSKI, CONDUCTOR
Severance Hall
11001 EUCLID AVENUE
CLEVELAND

C. J. VOSBURGH
MANAGER

July 11, 1942

White Goat Farm
Stockbridge, Mass.

Miss Nadia Reisenberg
Colonial Golf Club
Mount Kisco, New York

Dear Nadia:

I am ashamed that I didn't answer your letter sooner, but you know when summer comes you run around all day and hate to sit at a desk and write even to such a sweet nice girl as you are.

To tell you the truth, I have a whole bunch of scores which I haven't yet opened, and among them must be "Bright Land" by Mr. Triggs. As soon as I make up my mind to open them and inspect them, I will let you and the composer know.

How about coming sometime in September to Stockbridge? With love,

Yours,

Artur Rodzinski

AR/ef

That first concert went so well, Rodzinski seemed happy, I was elated, and it was a very exciting moment in my life. This was where it started, this was the first appearance, of which there would be so many hundreds and hundreds of others.

When we met again many years later and talked about our debuts, he said, "I wonder what you knew about playing at that time—I certainly didn't know much about conducting!"

NR, 1978

12 BUY WAR BONDS AND STAMPS ***Carnegie Hall Program***

Nadia Reisenberg, last soloist of the Philharmonic-Symphony subscription season, who plays the Prokofieff Piano Concerto No. 3 on April 13 and 14, reminisces with conductor Artur Rodzinski.

It was in Warsaw, the capital of Poland, in the year 1921. The Warsaw Philharmonic Orchestra was giving one of its regular performances. On the podium was a young conductor by the name of Artur Rodzinski. Among the members of the Orchestra were two young Russian refugees: Mischa Mischakoff, the concert-master, and Gregor Piatigorsky, the first 'cellist. A slender girl in her early teens appeared as a soloist and played the C-sharp minor Piano Concerto by Rimsky-Korsakoff. Her name was Nadia Reisenberg, another Russian refugee.

Many years later, Miss Reisenberg met Artur Rodzinski in New York. "Do you remember my appearance with you in Warsaw, Mr. Rodzinski?" asked Miss Reisenberg. "That was my first appearance with any orchestra, and I was so nervous."

"Do I remember?" answered Artur Rodzinski. "That was also my first appearance with the orchestra and I, too, was nervous!"

Twenty-three years have elapsed since that concert in Warsaw. Nadia Reisenberg is again appearing under the direction of Artur Rodzinski, this time playing the Prokofieff Piano Concerto No. 3, with the Philharmonic-Symphony Orchestra on April 13 and 14.

As to the Piano Concerto by Rimsky-Korsakoff, it was performed only once in New York, on March 24th, 1924, at Aeolian Hall, by the New York Symphony Orchestra, under the baton of Walter Damrosch. The soloist was Nadia Reisenberg.

New York Times, 4/14/44

RODZINSKI OFFERS PROKOFIEFF MUSIC

Third Piano Concerto Played by Philharmonic—Nadia Reisenberg Is Soloist

By NOEL STRAUS

The Philharmonic-Symphony Orchestra, under Artur Rodzinski, gave its final Thursday concert of the season last night in Carnegie Hall. For the occasion Dr. Rodzinski had devised a program featuring the Third piano concerto of Prokofieff, with Nadia Reisenberg as soloist, and also containing the First symphony of Beethoven and the First symphony of Brahms.

Not only Miss Reisenberg but the conductor and the orchestra as well were to be complimented on the splendid performance of the Prokofieff concerto, which completely realized the intentions of the exacting display piece. The rhythmic definition, crispness and clarity asked in every phrase of the work were fully at Miss Reisenberg's command. With ease and precision she conquered the formidable technical difficulties of the composition, which make all the greater demands on the soloist because there is almost no pause in the piano part throughout each of the three lengthy movements.

Technique is Praised

Miss Reisenberg possessed the necessary strength to cope with the more powerful pages of the score, and yet could invest the first and fourth variations of the slow division with all the needed limpidity and delicacy of tone, proving equally successful as colorist where bizarre sonorities were required. The opening theme of the initial allegro was admirable in its verve and sharp accentuation; the weighty chords leading to the second theme had monumental forcefulness, while the wild hurly-burly of the restatement was projected with masculine vigor and remarkable intensity.

The contrasted character of the variants of the andante could hardly have been more clearly established. In the finale, the manner in which the grotesquerie of the transition to the central episode was handled, and the positiveness with which the whole work was led to a climax of brilliance in the coda, called for special praise.

Fine Orchestral Support

Dr. Rodzinski and his men provided magnificent support in the concerto, a creation in which the orchestra has an unusually complicated and colorful scoring for a work of its genre.

Nadia Reisenberg, Pianist

Abresch

Soloist with the Philharmonic-Symphony Orchestra this afternoon

The first time I heard Nadia was in the Prokofiev Third, with the New York Philharmonic. She was one of the very glamorous women on the concert stage in those days—in addition to the great beauty she produced at the keyboard, she cut such a striking figure that we always enjoyed seeing as well as hearing her.

David Rubin, Steinway & Sons, 1974

The Prokofiev Third was the most remarkable performance, in the sense that it went as though we had had ten rehearsals. Rodzinski was just superb. The most important thing about him was his love for music, his enthusiasm for it. I don't think he knew what boredom meant, he had this tremendous involvement in everything he did.

NR, 1978

New York World Telegram, 10/12/45

Music

By Louis Biancolli

Kabalovsky's 2nd Concerto Is Brilliantly Played

Dmitri Kabalevsky's bright and rollicking Piano Concerto No. 2, premiered by the NBC two years ago, slipped into the local concert repertory at Carnegie Hall last night in a brilliant reading by Nadia Reisenberg and the Philharmonic-Symphony.

With Artur Rodzinski conducting, soloist and orchestra joined in a compact and warmly moulded rendering of one of the clearest symphonic communiques to come out of Russia. The language seemed so plain and lucid this time even a Foreign Minister couldn't go wrong working out the meaning.

Miss Reisenberg stood out again as one of the country's top keyboard gifts. The playing was strong, subtle, lyrical—as and when required. Tone surged from the piano in heaving waves and quieted down to faint filigree. Between were worlds of nuance. Some even thought Kabalevsky was getting more than he deserved.

There are fast, headlong stretches in this concerto calling for virtuoso powers. Miss Reisenberg had them, and more. Also there are soft nostalgic weavings, like faraway incantations, asking a poet's soul. The pianist showed that—and a heart, too. Put them all together and they spelt ovation.

This is an easy concerto to like. The themes make themselves at home in every part of the orchestra. Don't look for daring here in idiom or form. Kabalevsky will yet do better and newer things.

If more renderings like Miss Reisenberg's—or approaching it—are heard, the concerto ought to sweep like a streak along America's music lanes. The appeal is quick and human, even theatrical. You can't help liking it—unless you're a killjoy.

If Mr. Rodzinski and the orchestra didn't like it last night they concealed their feelings perfectly. What came through was zest and fervor, plus something very much like mass enjoyment. There was that kind of crispness in the tone.

The program started with what many view as the Mt. Everest of music—Beethoven's Fifth Symphony. Conductor and orchestra seemed to agree on the altitude. Whatever Beethoven had in mind here, it must have been big and shattering. The performance was in keeping.

Beethoven's C minor is an epochal moment in human conscience, like Dante's "Inferno" or Shakespeare's "Hamlet," or the atom bomb. What sings here is humanity at bay. A vision of evil and grandeur unfolds. You hear the shrieks of struggle—and the final victory. The finale ought to be read like the Bill of Rights. It was last night.

Also listed was the late Ernest Schelling's "A Victory Ball"—written in 1922 "to the memory of an American Soldier." The music grew out of a reading of Alfred Noyes' grim lines: "God! How the men grin by the wall, watching the fun at the Victory Ball." The harsh mood and shrill confusion of this odd fantasy still hold good.

More brash cynicism awaited the audience in Richard Strauss's fiendishly clever "Till Eulenspiegel." There at least there was time for comedy, too. I wonder what kind of music Strauss would write now?

MONDAY EVE., FEBRUARY 14, 1944 AT 8:30 O'CLOCK

Program

Pièces de Clavecin . . . . RAMEAU

Le Rappel des Oiseaux—1er Rigaudon—
2me Rigaudon—Double du 2me Rigaudon—
Musette en Rondeau—Tambourin

Sonata in B flat major, No. 10 (Posth.) . . . . . SCHUBERT

Molto moderato
Andante sostenuto
Scherzo, Allegro vivace con delicatezza
Allegro ma non troppo

Intermission

Third Sonata in A minor, op. 28 . . . . . PROKOFIEFF

Au Village, op. 40 } . . . . TSCHAIKOVSKY
Berceuse, op. 16, No. 1 }

Prelude, op. 23, No. 3 } . . . . RACHMANINOFF
Prelude, op. 32, No. 12 }

Toccata . . . . KHATCHATOURIAN

•

Steinway Piano

•

Tickets: $2.20, $1.65, $1.10, 83c; Loges, seating six, $19.80

New York Times, n.d.

MISS REISENBERG IN PIANO RECITAL

Russian-Born Artist Gives an Impressive Performance at Town Hall

Nadia Reisenberg, Russian-born pianist, gave a recital last night in Town Hall before a capacity audience which was most enthusiastic in its acclaim of her unusual art. Her program consisted of a group by Rameau, the posthumous B-flat major Sonata of Schubert, Opus 10, and works of Prokofieff, Tchaikovsky, Rachmaninoff and Khatchatourian.

Miss Reisenberg, who studied at the old Imperial Conservatory in St. Petersburg and at the age of 16 appeared as soloist with the Warsaw Philharmonic Orchestra, then directed by Artur Rodzinski, and later studied here with Alexander Lambert and Josef Hofmann, is widely regarded as one of the outstanding women pianists of the day. As a bravura player, she must be ranked among the best of her sex.

Her great technical command of the keyboard was shown last night throughout the program, especially in Prokofieff's Third Sonata in A minor, Opus 28, in the Rachmaninoff Prelude, Opus 23, No. 2, and in the closing Toccata of Khatchatourian. These works require great brilliance and power, and Miss Reisenberg fully supplied it. Her tone was admirable, with lovely shading and a discriminating use of color.

It could be big, as in the Khatchatourian Toccata, where, however, the strings occasionally jangled, and it could be delicate and even feathery, as in the last two movements of the Schubert Sonata, particularly in the third, Scherzo, Allegro vivace con delicatezza.

Above and below all of this pianism, Miss Reisenberg disclosed a sound musicianship and a poetic feeling which, added to the technical ability to set forth the general form of a composition while still bringing out clearly all of its contrasting parts and distinct voices, resulted in piano playing of a high artistic value.

Some of the finest playing of the evening appeared in the half-dozen encores given at the end of the printed list, beginning with Chopin and ending with a Scriabin étude.

R. L.

Family Interlude III

BLESSED EVENT(S): Father—Tommy (Schuster); Mother—Bonnie Lassie (Sherman); Children—Do, Re, Mi, Fa, Sol, La, Ti

NR and Sasha

Alex and NR

NR and Sasha

Pianist Raises Her Family, Too

Trick Is Having Right Husband, Says She

By LOUIS BIANCOLLI.

Bringing up a family and being one of the world's top concert pianists may seem to rule one another out. Especially when the keyboard virtuoso is of the female gender.

Not with Nadia Reisenberg, though. She has combined both, plus a heavy teaching schedule, and has been an artist at it, too. It has meant a life of self-discipline and an exactly adjusted family life. First requisite, she feels, is having the right husband.

"It has to begin with his attitude," said the handsome Russo-American concert star, who returns to Town Hall Monday with a sheaf of piano classics.

"Without complete co-operation and understanding on the husband's part it would be utterly impossible. My husband has always encouraged and respected—and understood.

"When I have a concert he acts like a sentinel—sees that I am free from all other duties and outside disturbances."

The second requisite is having the right children. There are two in the Reisenberg home—Alexander, 19, now a seaman 1C, and Robert, 12.

Children Co-operate, Too.

"Like my husband, the boys always know when my privacy and work are not to be disturbed," she said. "The respect my husband has always shown has passed on to them.

"Of course, I would like to give them more time. We are such good pals. Associating with them more would probably be good for all of us.

"But there is one redeeming feature about this: not being with you too much, they don't take you for granted. They don't consider you a piece of furniture.

"My children, for example, make appointments with me. My big boy will call me up and say 'When can we have lunch together, mother?' They learned very fast to take care of themselves.

"Neither of them looks like a gangster. They're quite gentle, as a matter of fact, and I would say that in spite of my career, which often takes me to distant places, they are rather well brought up.

"But it all depends on the husband. It would have been out of the question in my case if he hadn't understood that music was too much a part of my life, that it was like eating and breathing—something I couldn't put away."

Mme. Reisenberg's husband is I. J. Sherman, a brilliant economist with a Heidelberg doctorate. His interest in music was academic till he met his wife. Then the fever caught him, too.

"When he saw that music was something I couldn't live without he felt it time to learn more about it," she said. "Now he is really an expert in musical literature."

Boys Good Musicians.

Naturally the boys have also absorbed some of their mother's passion for music. Both are good musicians and play the piano quite well, but Mme. Reisenberg isn't counting on their turning out concert virtuosos.

"One in the family is enough," she said. 'Besides, neither seems gifted the way a concert pianist should be. Talent should be of the very first caliber before deciding on a career.

Featured on Monday's Town Hall program is Schumann's F-sharp minor sonata. How it got there is a neat little story linking the names of Mozart, Schubert, Clara Schumann and Nadia Reisenberg.

As a reward for playing Schubert's B flat major sonata uncut last season, John Bass, noted music manuscript collector, gave her three items—a program of Mrs. Schumann's showing the Schubert listing, a photo of the Schumann couple, and Mrs. Schumann's letter to a baroness.

"Before I had a chance to read the contents of the letter," said Mme. Reisenberg, "I remarked to Mr. Bass how thrilling it was to hold a letter of Clara Schumann's and that in just a few moments I would be, so to speak, looking into a corner of her soul.

"Imagine my amazement when I read the letter through and found it to be a note of thanks for a gift made to her by the baroness. And what was the gift? A letter of Mozart's. And there was Clara Schumann writing how touched she was having the feeling that she had 'entered Mozart's soul!'

"I felt then as though I was in direct contact with Mozart via Clara Schumann."

Bob, Sasha, NR, and Alex

Sasha, NR, Alex, Ruth (Alex's wife), and Bob

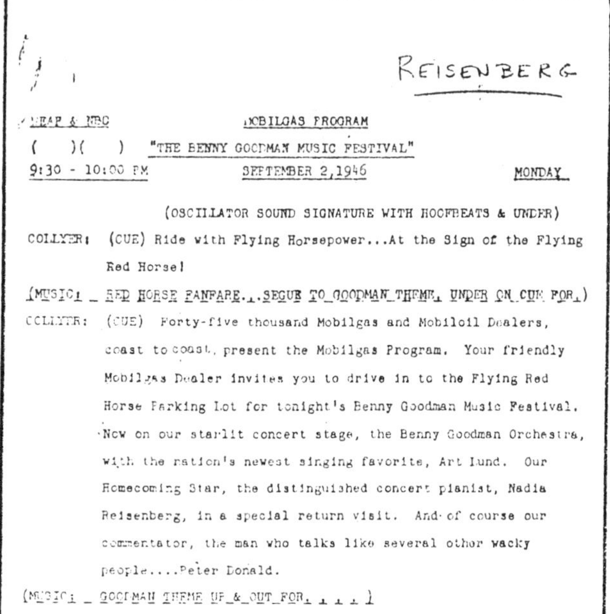

REISENBERG

WEAF & NBC — MOBILGAS PROGRAM

( )( ) "THE BENNY GOODMAN MUSIC FESTIVAL"

9:30 - 10:00 PM — SEPTEMBER 2,1946 — MONDAY

(OSCILLATOR SOUND SIGNATURE WITH HOOFBEATS & UNDER)

COLLYER: (CUE) Ride with Flying Horsepower...At the Sign of the Flying Red Horse!

(MUSIC: RED HORSE FANFARE...SEGUE TO GOODMAN THEME, UNDER ON CUE FOR.)

COLLYER: (CUE) Forty-five thousand Mobilgas and Mobiloil Dealers, coast to coast, present the Mobilgas Program. Your friendly Mobilgas Dealer invites you to drive in to the Flying Red Horse Parking Lot for tonight's Benny Goodman Music Festival. Now on our starlit concert stage, the Benny Goodman Orchestra, with the nation's newest singing favorite, Art Lund. Our Homecoming Star, the distinguished concert pianist, Nadia Reisenberg, in a special return visit. And of course our commentator, the man who talks like several other wacky people....Peter Donald.

(MUSIC: GOODMAN THEME UP & OUT FOR...)

COLLYER: Now a happy holiday hello from the head man himself -- Benny Goodman!

(APPLAUSE)

-6-

GOODMAN: Just ten weeks ago tonight we began this concert series with a feature called "Homecoming Time". During past weeks it's been our pleasure to bring back outstanding performers who started with us or who have been our associates in the past. Among our first Homecoming stars was one of the world's great pianists, and we were very happy when she accepted our invitation to pay us a return visit. We are proud to present her now as the star of this evening's festival -- Miss Nadia Reisenberg!

(APPLAUSE)

REISENBERG: Thank you, Benny. I so enjoyed my first visit, and I AM very glad to be back. But since you and I have been rehearsing a piano-clarinet work all week, don't you think you should tell our friends about it,

GOODMAN: We'd like to save that for later. Right now we want our friends to hear one of the perfect combinations in music.... a chopin work played by a truly great artist. Miss Nadia Reisenberg plays Chopin's "Waltz in A flat major."

(MUSIC: "WALTZ IN A FLAT MAJOR"......NADIA REISENBERG....)

(APPLAUSE)

GOODMAN SHOW — 10 — REVISED

GOODMAN: Nadia Reisenberg has returned; and as she told you we have been rehearsing all week on a work for piano and clarinet. While I walk over to the piano let's get Mr. Collyer to tell you all about it.

COLLYER: The work began as a lilting melody by Mozart in the popular opera "Don Giovanni". The great Beethoven wrote a set of variations on the theme and this has been transcribed for piano and clarinet by Simeon Bellison - solo clarinetist with the New York Philharmonic Symphony So here are Nadia Reisenberg and Benny Goodman in Beethoven's variations on a theme from "Don Giovanni".

(MUSIC: ..."DON GIOVANNI"...THEME...REISENBERG & GOODMAN....)

(APPLAUSE)

Nadia Reisenberg was a dedicated musician and altogether charming person. We recorded the Brahms Eb Sonata for Columbia and another unforgettable musical get-together took place in California. She was a close friend of Gregor Piatigorsky and invited him to come to my house in Westwood Village to play the Brahms Trio. The music making was a joy, their conversations were scintillating—she was a great lady.

Benny Goodman, 1985

I remember Nadia with fond admiration and esteem, both as an artist and a person. Hers was a distinctive contribution to the musical scene.
Morton Gould, 1985

CARNEGIE HALL ◆ FRI. EVE., NOVEMBER 21, 1947 ◆ AT 8:30 O'CLOCK

NADIA REISENBERG

Pianist

PROGRAM

I

Suite in G Minor . . . . HANDEL

II

Sonata in A Minor, K. 310 . . . . MOZART

Rondo Brillante . . . . WEBER

III

Sonata in B Minor . . . . CHOPIN

Intermission

IV

Four Excursions . . . . BARBER

V

Six Etudes . . . . SCRIABINE

Etude in F Sharp Major . . . . . . STRAVINSKY

Etude in D Minor . . . . PROKOFIEFF

Steinway Piano

Tickets: Box Seats $3.00 and $2.40 (Boxes Seat 8); Dress Circle $1.80;
Orchestra $2.40 and $1.80; Balcony $1.20 (Tax Incl.)
On Sale at Carnegie Hall Box Office
Concert Management: RAY HALMANS, 119 West 57th Street, New York, 19

Excursions

I

Samuel Barber, Op. 20

Un poco allegro ♩= 144

Piano

p

senza pedale

poco f

con pedale

senza pedale

poco f

p

Samuel Barber and NR

40969 c

Georges Enesco

Georges Enesco, Richard Tucker, and NR

CARNEGIE HALL
Season 1948-1949

Wednesday Evening, November 24th, at 8:30 o'clock

BENEFIT CONCERT
for the JEWISH CHILDREN OF ISRAEL

sponsored by
THE UNITED ROMANIAN JEWS OF AMERICA
Dr. Charles H. Kremer, Chairman, Concert Committee

GEORGES ENESCO *Violinist* — RICHARD TUCKER *Tenor*

NADIA REISENBERG *Pianist* — PIA IGY *Soprano*

PROGRAM

I.

Sonata in A major . . . . *César Franck*
Allegretto ben moderato
Allegro
Recitative-Fantasia
Allegretto poco mosso
GEORGES ENESCO and NADIA REISENBERG

II.

Sound an Alarm . . . . *George Frederick Handel*
(Recitative and Aria from "Judas Maccabeus")
Sound an alarm! Sound an alarm, your silver trumpets sound,
And call the brave, and only brave, and only brave around.
Who listeth, follow! to the field again!
Justice with courage is a thousand men, is a thousand men.
Sound an alarm!

Program Continued on Second Page Following

Georges Enesco and NR

A Busman's Holiday: Tanglewood, Summer 1948

Seymour Lipkin, Freda Berkowitz, Alexander Schneider, NR, and Ralph Berkowitz

From the days I first knew her in the '30s, Nadia could not be indifferent to things musical—her nature was for enthusiasms, for zealous interests. One's most sincere accolade for her ceaseless warmth as a teacher and as a magnificent pianist is that she was a musician's musician...

Ralph Berkowitz, 1985

I was always struck by the peculiar intensity of the feelings Nadia's students had for her. Their special relationship reflected her intense concern for her as well as their incredible devotion to her.

Seymour Lipkin, 1985

Oh, how I loved playing four-hands with Nadia! She was such a wonderful pianist...and such a charming woman!

Boris Goldovsky, 1985

Ralph Berkowitz, NR, and William Primrose

NR and Gregor Piatigorsky

NR and Boris Goldovsky

Gregor Piatigorsky, NR, and Jascha Heifetz

NR and Leonard Bernstein

Abram Chasins, NR, and Ralph Berkowitz

No one has contributed more to our beloved art than you, my dear, valued friend Nadia . . .
Abram Chasins, 1957

The 1950s and '60s

PITTSBURGH

NEW FRIENDS OF MUSIC, INC.

(a non-profit organization)

Twelfth Season, 1949-1950

BUDAPEST STRING QUARTET

Josef Roisman, *violin*

Jac Gorodetsky, *violin*

Boris Kroyt, *viola*

Mischa Schneider, *violoncello*

NADIA REISENBERG, *piano*

PHILIP SKLAR, *double bass*

CARNEGIE MUSIC HALL

MONDAY EVENING, FEBRUARY 13, 1950 AT EIGHT-THIRTY

PROGRAM FOR NEXT CONCERT

AT TOWN HALL — NOVEMBER 13, 1950

SECOND PROGRAM

Soloists: **Nadia Reisenberg, *Piano***
Joseph Fuchs, *Violin*
Leonard Rose, *'Cello*

NADIA REISENBERG

JOSEPH FUCHS

LEONARD ROSE

1. Symphonic Essay . . . . *Robert Nagel*
   First Performance

2. Violin Concerto No. 4 in D major, k. 218 . . . *W. A. Mozart*
   Allegro
   Andante Cantabile
   Rondeau. Andante Grazioso—Allegro ma non troppo
   SOLOIST: Joseph Fuchs

3. Piano Concerto, op. 30 . . . . *Rimsky-Korsakoff*
   Moderato—Allegretto quasi polacca—Andante mosso—Allegro
   SOLOIST: NADIA REISENBERG

INTERMISSION

4. 'Cello Concerto in B flat major . . . . . . *Luigi Boccherini*
   Allegro Moderato
   Adagio
   Rondo—Allegro
   SOLOIST: Leonard Rose

5. Concerto for piano, violin, violoncello and orchestra, op. 56 . . . . *Alfredo Casella*
   First New York Performance
   Largo, ampio, solenne—Allegro molto vivace
   Adagio
   Rondo—tempo de giga—allegro vivace ma non troppo
   SOLOISTS: Nadia Reisenberg, Joseph Fuchs, Leonard Rose

New York Herald Tribune, 11/14/50

MUSIC

By VIRGIL THOMSON

LITTLE ORCHESTRA SOCIETY

TOWN HALL

Thomas Scherman, conductor, second concert of the season last night. Soloists, Nadia Reisenberg, pianist; Joseph Fuchs, violinist; Leonard Rose, cellist. The program:

Symphonic Essay....................Robert Nagel
(First performance)
Violin Concerto No. 4, in D major (K. 216), Mozart
Piano Concerto, Op. 30......Rimsky-Korsakoff
Cello Concerto in B flat major......Boccherini
Concerto for piano, violin and cello, Op. 56, Casella
(First New York performance)

Virtuosity, Musicianship, New Pieces

THREE first-class soloists and two pieces new to New York were the offering last night of the Little Orchestra Society, playing its second concert of the season in Town Hall. Nadia Reisenberg, Joseph Fuchs and Leonard Rose, each alone, would have assured a fine evening. All three together made a wealth of beautiful playing, framed by the support of Thomas Scherman and his good orchestra.

* * *

The newer of the two new pieces was a "Symphonic Essay" by Robert Nagel, twenty-six-year-old trumpet player. This contrapuntally conceived work, all made of long lines harmoniously enjoying thematic imitation, might have passed for student work, for all its taste and skill, had not the melodic contours of the work been so deeply marked by grace and the expression of the whole so intense. Mr. Nagel has an instinct for shapeliness and a concentration of feeling that show him to be a composer of no ordinary integrity. His Essay has character and strength and a luminous inner life that do not seem to be imitated, could not be, from the writing of other composers, though his technical methods are in no way novel.

Alfredo Casella's Concerto for Piano, Violin, Violincello and Orchestra, Opus 56, though composed in 1933, is new to New York. It is a work of master craftsmanship, eclectic, neoclassic, agreeable, not very original in thought. It is a bit insistent rhythmically and not strong as to tunes, but it sounds good at all times. It is complex of texture, rich of harmony, contrapuntally busy, instrumentally brilliant, soundly musical throughout and not lacking in audience appeal. What it lacks chiefly is urgency. It could exist or not, and the history of music would be just the same.

* * *

Thomas Scherman

Helen Merrill

Conductor of the Little Orchestra Society

Rimsky-Korsakoff's Piano Concerto, Opus 30, is a one-movement piece, an olio of Russian folktunes. It has a faded Victorian charm but not much coherence. It was chiefly a pleasure for Miss Reisenburg's excellent piano-playing, ever clean and strong and intelligent.

New York Times, 11/14/50

LITTLE ORCHESTRA PLAYS 4 CONCERTOS

Nagle's New Symphonic Piece Also on Program—3 Soloists Heard During Evening

Four concertos and a new symphonic piece by Robert Nagel, 26-year-old American composer, made up the characteristically generous program offered at Town Hall last night by Thomas Scherman and his Little Orchestra Society. For the concertos there were three different soloists, who had a concerto each and who then joined in the first New York performance of a triple concerto by the late Alfredo Casella.

There was some superficial music on this program, some that was lively, some that was deep and affecting, but nothing out and out dull. Take Rimsky-Korsakoff's Piano Concerto, Op. 30—his only concerto—which had brilliant, virtuoso work by Nadia Raisenberg, the soloist. Built on a plaintive, folk theme, it is, by the composer's own admission, meant to follow in the footsteps of the Liszt concertos. It is colorful, noisy and brief. You don't get bored listening to it, even if you don't look forward to hearing it often. Give Miss Reisenberg —and Mr. Scherman—credit for bringing enormous gusto to it.

In the Mozart Violin Concerto in D (K. 218) the soloist was Joseph Fuchs. His playing had felicity of style. The tone was light and airy, the phrasing graceful and searching. The work is familiar, but Mr. Fuchs played it with freshness of feeling.

Leonard Rose, 'cellist, who was the third soloist, appeared in Boccherini's B-flat major Concerto, and almost walked off with the show. His playing of the sustained song of the slow movement was rich in tone, and sensitive in nuance, the work of a first-rate musician.

The three soloists did a workmanlike job in the Casella concerto, which is eclectic music, now perky and forward-looking, and now sober and conservative. Whether Casella made the most effective use of his three soloists is open to question, but he knew how to keep an orchestra busy. Mr. Scherman did a good job keeping all his elements busy during the enfolding of the piece.

Mr. Nagel's "Symphonic Essay" turned out to be a modest, serious piece. Its ideas and their working out are not exactly novel, but one must respect a composer—he is also first trumpet of this orchestra—who is not afraid to be a little old-fashioned if that is how he feels. H. T.

New York University Glee Club

ALFRED M. GREENFIELD, Director

Twenty-second Annual

TOWN HALL CONCERT

Friday, December 7th, 1951

at Eight-Thirty o'clock

NADIA REISENBERG, Pianist
Assisting Artist

Bobby's mistake as a child was that he studied with me. He should have studied with another teacher who would have made him practice. Now he's the one who tells me, "Why *didn't* you make me practice?" It's the old refrain...

NR, 1977

Perhaps my children have missed something (because of my career) but on the other hand they look forward to having lunch with me and going somewhere with me. They don't take so much for granted.

NR, 1943

New York Times, 12/7/51

PIANO DUET FEATURES N. Y. U. CLUB CONCERT

A piano duet by Nadia Reisenberg, internationally known pianist, and her son, Robert Sherman, will highlight the twenty-second annual Town Hall Concert of the New York University Glee Club at 8:30 tonight. The performance will mark the first time Miss Reisenberg and Mr. Sherman have appeared together professionally.

A senior at New York University's College of Arts and Science, Mr. Sherman, who is nineteen years old, is opposed to becoming a professional concert artist. "I'm really not that good on the piano," he said yesterday at rehearsal. "I play mostly for my own amusement and get serious only for a special engagement."

Miss Reisenberg, who uses her maiden name professionally, received her early musical education at the Imperial Conservatory in St. Petersburg. In her early teens, she was soloist with the Warsaw Philharmonic Orchestra, and has played all twenty-seven piano concertos by Mozart with the WOR Symphony Orchestra.

Miss Reisenberg will play tonight Mozart's "Pastorale Variée" and "Rhapsodie Espagnole" by Liszt, and will team with her son to play three of the movements of Darius Milhaud's "Scaramouche."

Teaching at the University of Southern California

California Judging Panel: Frank Sheridan, pianist; NR; and Raymond Kendall, Music Dean at the University of Southern California

Well, as Rosalie Talbott used to say in Tulsa, I "dood" it again. The concert was an enormous success, the hall filled to the brim, plus about 50 standees (who after intermission were seated on the stage). [Alfred] Wallenstein, [Raymond] Kendall and all who came backstage could not have seemed more enthusiastic, and all told me that I never played better. I did not think so. Somehow I did not enjoy playing as much as I hoped I would, and I know that I can play better. You know me—if I think I can do better, no praise can alter my opinion.

NR, 1952

THE NEW YORK TIMES, SUNDAY, OCTOBER 11, 1953.

STEINWAY CENTENARY WILL BE OBSERVED WITH GALA CONCERT NOV. 19

The New York Times (by Sam Falk)

There will be thirty pianists taking part in the Carnegie Hall program in addition to the New York Philharmonic. Here are seventeen of the group. Starting at lower left and moving clockwise, they are Ania Dorfmann, Moura Lympany, Menahem Pressler, Muriel Kerr, Eugene List, Alexander Uninsky, Beveridge Webster, Sascha Gorodnitzki, Gyorgy Sandor, Leonid Hambro, Erno Balogh, Gary Graffman, Abram Chasins, Nadia Reisenberg, Franz Rupp, Jacques Abram and Constance Keene.

To Nadia Reisenberg

Charles G. Steinway

Theodore D. Steinway

Frederick Steinway

H. Steinweg
Instrumentenmacher
in
SEESEN

The oldest known Steinway piano and its label. Built by Henry Engelhard Steinway in 1836, 17 years before the founding of Steinway & Sons.

JANUARY 14, 1953

All Beethoven Program

I. Sonata II, op. 5, No. 2 in G minor

II. Sonata IV, op. 102, No. 1 in C minor

III. Twelve Variations on "Ein Maedchen oder Weibchen" from the opera: "Magic Flute", by Mozart

JANUARY 21, 1953

All Beethoven Program

I. Sonata I, op. 5, No. 1 in F major

II. Seven Variations on the duet "Bei Maennern, welche Liebe fuehlen" from the opera: "Magic Flute", by Mozart

III. Twelve Variations on a theme from "Judas Maccabaeus" by Handel

IV. Sonata V, op. 10, No. 2 in D major

JANUARY 28, 1953

I. Sonata No. 1, op. 38 in E minor .... BRAHMS

II. Sonata, op. 65 .... CHOPIN

III. Sonata, op. 19 .... RACHMANINOFF

Steinway Piano

Subscription: $3.60, $4.80, $6.00, including tax
Single Admission: $1.50, $1.80, $2.40, including tax

For reservations call TRafalger 6-2221, or mail check to YM-YWHA, Lexington Avenue at 92nd Street, New York 28, N. Y.

New York Herald Tribune, 1/29/53

Schuster-Reisenberg Program

Joseph Schuster and Nadia Reisenberg completed their concert series of music for cello and piano last night in the Kaufmann Auditorium of the Y. M. and Y. W. H. A. at Lexington Ave. and 92d St. The first two programs had been devoted to sonatas and variations by Beethoven; yesterday's closing list offered three sonatas of differing dates and nationalities, but all romantic in character, those by Brahms in E minor, Op. 38, No. 1; Chopin, Op. 65, and Rachmaninoff, Op. 19.

Both artists were in admirable technical form. Mr. Schuster's cello tone was even in texture and appealing in quality; its inherent warmth increased as the program advanced, and was particularly notable in the melodically opulent, rather discursive Rachmaninoff work.

While his playing gave an impression of interpretative understanding, Miss Reisenberg's had a wider expressive range; its climaxes were outspoken, but the pianist also observed dynamic discretion, and the interpretations as a whole were mutually sympathetic and tonally well balanced.

F. D. P.

Family Interlude IV

Top: Clara, Bob Arlan, Zena Arlan (NR's cousin and Ruth Greenberg's sister). Bottom: Ruth Greenberg (NR's cousin), and NR.

Q. Sometime in the early 1950s, you stopped most of your touring, gave up solo recitals, and concentrated much more intensively on teaching and playing chamber music. Was this a conscious decision, and in retrospect, are you happy with it?

A. I definitely was happy about it. I'm not one of those people who loves to stay overnight each day in another city. I used to be very homesick, I'd miss the children. I actually had absolutely enough of all of it. The glamor of it didn't interest me anymore. I'd had it all, and I thik in its own time this was wonderful. But since chamber music, as I've told you many times before, is my greatest love, I santed more time to pursue it. Besides, I wanted to teach more. I think I really have as much talent for teaching as I have for performance (yes, you have to have talent for that as well), and it has given me so much satisfaction, because the success of my students is vicariously my own success. I really am enjoying both chamber playing and teaching, so I'm quite content my career took that turn.

NR, 1978

At Alex and Ruth's Wedding, 1950

At Bob and Ruth's Wedding, 1956

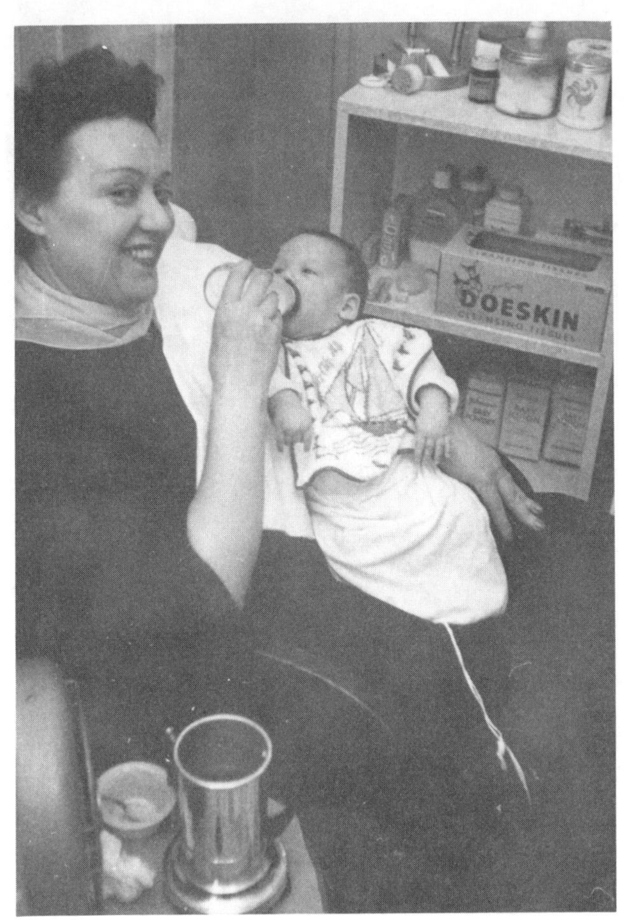

NR with first grandson David (born November 28, 1956)

IN OSSINING, NEW YORK

NR with Jenny Grey (Zwibak)

NR and Newta

NR with William Horne

Mariasha Boritch and NR

Artists can be selfish—sometimes they have to be the center of the world—but Nadia was different. She was very much interested in other people, and shared their lives. She was a complete friend.

Andrei Sedych, 1985

He (Sasha) was not only her husband but her most ardent admirer, as well as adviser, critic, manager, and true friend.

Andrei Sedych, 1955

Last photo of Sasha (with Bob and NR), Fort Dix, New Jersey, 6/6/55

J. ANTHONY MARCUS
SCARSDALE, N. Y.

Aug. 8, 1955.

My dear Mrs. Sherman:

Like many other friends of your late husband, I was deeply moved by his sudden passing. A few days before I left on a business trip to Southwestern states, I ran into Isaac on 42nd Street. I was always thrilled to see him. There was something so genuine, so sincere, and so friendly about him, and this has been the case ever since my first meeting with him in the very early twenties.

Then in March, 1926 I happened to be in Odessa. The day after I landed in the London Hotel there appeared a big interview with me in the Odesskiya Izvestiya. That afternoon the room clerk telephoned to me that an elderly man wanted to see me. I told him to bring him up. A neat-looking, bearded gentleman came up and was quite not at ease, I noticed. I tried to make him feel at home, asked for some tea in the usual Russian fashion and before long he was full of smiles.

"Do you by any chance know my son, Isaac, in New York?" It was, of course, quite inadequate information. I asked for more about his son. And to my great surprise I was sitting in front of my friend's father. He had read the interview and took a chance to ascertain if I happened to know his son. He had a very serious message he wanted me to bring to New York. He wanted to go to Palestine, he told me. There was no life for him in Russia. I should ask Isaac to try to get him out of there.

Naturally one of the first things I did when I returned to New York was to see Isaac and deliver the message. I later learned that he had reached Palestine happily. What a wonderful thing to have done; wonderful for both--father and son.

We have all lost a loyal friend. Your loss, of course, is the greatest. Only those of us who had truly great loves and suffered their premature loss know what it means to the survivor. Only time can heal the wound. May time do its job for you quicker than it has done for some of us. There is nothing one can do for those who have passed on. There is much we can do for ourselves, as that is what the departed would want us to do. I therefore hope that you will be blessed with the best of health and cheer and continue the beautiful musical work with which you have enriched us all.

Sincerely yours,

Sasha,

Without you, I feel so empty,
I hurt so much.
Oh, darling, my dearest friend!
My tears just keep falling,
I want so much to be with you.

NR, August 11, 1955, 1 A.M.

It's as though everything is as it always was,
And there are no special changes here.
The apartment is clean and nice—
Why, then, do I fill the walls with my tears?
I get up as usual, I get dressed,
But my heart is oppressed in unbearable anguish.

I practise as before,
But my thoughts wander far away.
I am surrounded by love,
But is is your love I need.
They don't leave me alone, not for a moment,
But without you, I am alone all the same.

NR, August 11, 1955, 2 A.M.

Nadia was scheduled to record the Chopin Nocturnes shortly after Sasha died, and she refused to postpone the sessions. You can imagine the kind of emotional climate we were living in, but it drew us closer than ever before. In fact, I think that the release of such deep sorrow through her music may have saved her sanity.

Clara, 1985

I was with her every day, listening on the couch as she practised and cried, cried and practised. It was incredibly painful, and yet at the same time what an unforgettable experience to hear her pouring her very heart and soul into the music.

Newta, 1985

I never saw Nadia under such emotional strain before or since. It was heartbreaking. Her breathing became so heavy during the taping that I was afraid it would ruin the record, but if there was desperation, there was inspiration too. She never played anything more beautifully than those Chopin Nocturnes.

Clara, 1985

It is a little painful listening to my recordings again because I always feel that I could improve things, but basically I'm very satisfied. You see, I've always aimed for simplicity in my conceptions. I hate doing anything "original" for its own sake, and while I feel everything I play very deeply, the music always comes first. I have no right to meddle with it. As a result, many of the basic values are still there, even after twenty years. I might play certain things more simply now, or with greater refinement—after all, I'm a few years older!—but I would still try to get at the truth of the music, and that truth hasn't changed.

NR, 1974

Also on an extremely high level of Chopianistics is Nadia Reisenberg's three-disk performance of all the Chopin **Mazurkas** plus, on the last side, the **Barcarolle, Berceuse** and **Allegro de Concerto** (Westminster; the three disks are not in album form but available separately). Some of Chopin's most magical music is contained in the mazurkas, and Miss Reisenberg fully rises to the occasion. Her playing is neat without being academic, poetic without being flamboyant, intelligent without being stuffy. For the most part her conceptions are delicate; she does not have the big line that Rubinstein brought to his recording of the mazurkas. But, then again, Miss Reisenberg has some things Rubinstein did not have, including an entirely different type of rubato (and one just as convincing). She also gives more for the money, in that her three disks contain all but two of the posthumous mazurkas and three additional works—the "Barcarolle," "Berceuse" and seldom-heard "Allegro de Concerto."

Some of the big exhibitionistic pieces of the piano reper-

CAPSULE REVIEWS OF NEW RECORDINGS

BRAHMS: Two Sonatas for Viola and Piano; Paul Doktor and Nadia Reisenberg (Westminster). Several versions of these sonatas have been recorded in their version for clarinet and piano, but this is the first with the viola alternate. They receive splendid performances here—musicianly, with tempos that make sense and natural-sounding phrasing. First-class recorded sound, too.

RACHMANINOFF: Piano Pieces, Op. 3 and 10; Nadia Reisenberg, pianist (Westminster). Op. 3 is the set that contains Rachmaninoff's most popular piece, the Prelude in C sharp minor. Some of the others are almost as well known (and several are hardly ever heard). Miss Reisenberg plays with artistry and plenty of technique, and the recorded sound is excellent.

HARPERS MAGAZINE 9/58

Haydn: Sonatas for Piano, Vol. 2 (#52 in E Flat, #43 in A Flat, #34 in E Minor). Nadia Reisenberg. Westminster XWN 18358.

Miss Reisenberg is another brilliantly "masculine" pianist, who knows how to shape these subtle early piano sonatas to the proper size. They are as big as they ought to be—most pianists misinterpret their thin lines as a false delicacy—but Reisenberg never takes them out of scale, even so. The ever-moving E Flat Sonata, Haydn's last and composed in the fruitful period after the end of his symphonic writing, is a marvelous work of experimentation in Haydn's own somewhat mystical form of Romanticism. Its key relationships are extraordinary—the middle movement is in E major, musically poles away from the E Flat of the outer movements. It's good to have this sonata in such an intelligent and expressive playing and the same goes for the two earlier works. There'll be more to come.

Brahms: Sonatas for Viola and Piano, Opus 120 (Paul Doktor, viola; Nadia Reisenberg, piano; Westminster WN-18114: 12"). These were published simultaneously as sonatas for clarinet and for viola. In the latter version, I had not heard them until now. Whereafter, I agree with Sir Donald Tovey that they are better in the viola than in the clarinet version; there is a tenderness in the strings that the reed cannot quite match, even when Kell is the artist, and it is important here. Doktor and Reisenberg — a marvelously expressive pianist — leave nothing to be desired, and the reproduction is adequate to their intent.

I would like to congratulate you on your recording of "Masterpieces of Haydn." That is superb artistry and piano playing of the highest order.

Aldo Parisot

MOZART: ***Sonatas for Piano, Four Hands: in F, K. 497; in B flat, K. 358. Adagio and Allegro in F minor, K. 594***

Nadia Reisenberg. Artur Balsam, piano.
○ MUSICAL HERITAGE SOCIETY MHS 632. LP. $2.50.
○ ○ MUSICAL HERITAGE SOCIETY MHS 632. SD. $2.50.

If you can't find anyone to play these Sonatas with you, the next best thing is to listen to them in a recording like this one, where the artistry of the playing will make up in part for the ineffable satisfaction one gets when performing in a duet, even with an occasional stumble.

Miss Reisenberg and Mr. Balsam, each long and well known as a sensitive and poetic performer, make a fine team. There is precision enough, but it is never cold or mechanical, and the music is beautifully sung. K. 358, written when Mozart was about eighteen, is light and charming; K. 497 is a big work, with considerable brilliance in the fast movements and engrossing dialogues in the Andante. K. 594 was originally written for a mechanical organ; there is no basis in fact for the statement in the notes that the four-hand arrangement is by the composer, nor does the work come off very well in this guise. Some top notes are split, but otherwise the sound is good. N.B.

Haydn Album Great

RECORDS

Detriot Sunday Times, 7/20/58

By John Sweeney

Franz Joseph Haydn, who could combine in one single work lofty eloquence and the most outrageous musical puns, is possibly the most unassuming and beloved genius the world has known.

Affectionately known as "Papa" Haydn, he is recognized as the father of the modern symphony and chamber music, and the composer of some of music's greatest oratories.

HOWEVER, his piano works, which include some 52 sonatas and at least as many sets of variations and smaller works, have been studiously ignored by pianists and audiences.

Although not each of them shows Haydn at the full height of his powers, all of them give at least one aspect of his unique greatness, many are works of genius, and his last sonata, No. 52 in E flat major, must be rated a masterpiece.

This last sonata and two others, No. 34 in E minor and No. 43 in A flat major, comprise Volume 2 (Westminster XWN 18358, $3.98) in a new series devoted to Haydn's piano sonatas.

BEAUTIFULLY performed by Nadia Reisenberg, they are delightful examples of Haydn at his pianistic great- [illegible] of novelty besides containing immortal passages of grace, wit and noble lyricism.

Miss Reisenberg has a rare affinity for Haydn. Her touch is light but not superficial, her interpretations broad and yet supple, her technique clean and pearly.

SHE HAS an unfaltering understanding of classical form and spirit, and her playing, though distinctively feminine, is forceful and vivacious.

Carefully recorded with a brilliant, though never strident, tone, this album can be highly recommended for those looking for great, but not unhackneyed, music.

HAYDN
Smaller Pieces for Clavier

Andante varié in F minor; Arietta and Variations, in A; Arietta and Variations, in E-flat; Capriccio in G; Fantasy in C; Theme and Variations, in C.

Nadia Reisenberg, piano.
WESTMINSTER 18057. 12-in. $4.98 (or $3.98).

All are new to LP except the F minor Andante and Variations. The others have been excluded hitherto presumably in deference to the traditions of public performance. They do not roll enough thunder to make a big hall gape. But they are packed with fancy and invention allied to taste, and strew bright patterns in resourceful alterations lively with unexpected notions.

No other Haydn record of the solo piano strikes the memory as comparable to this one in the precise mirroring of small piano-sound, or in the unruffled, obstinate refinement of the feathered filigrees whirled into a classic geometry by Miss Reisenberg. Why have we had no other Haydn from her, and why no Mozart?

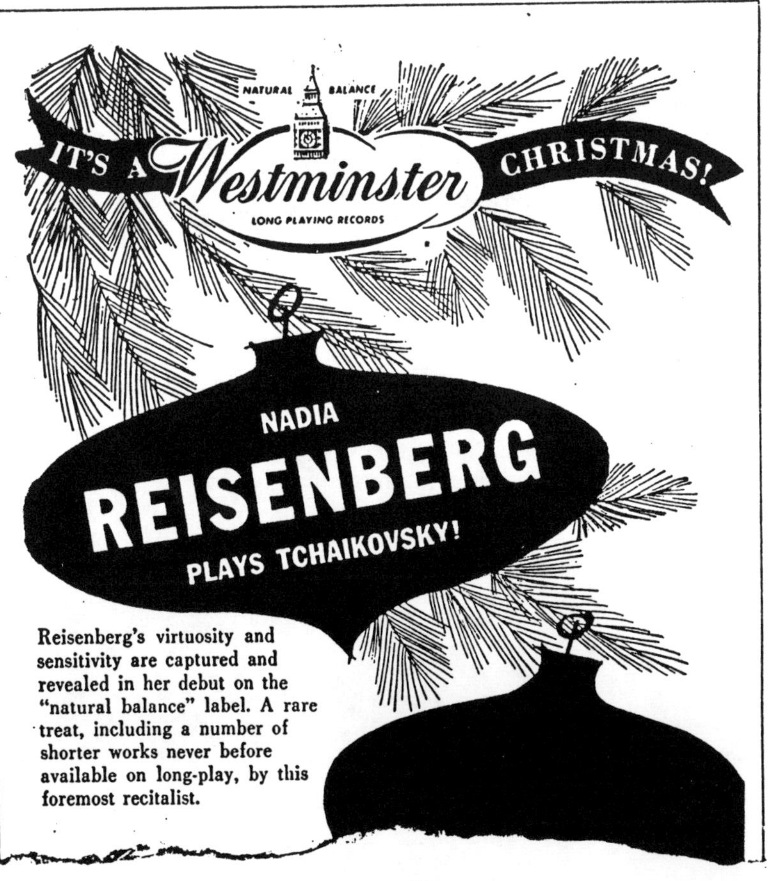

Just a few words about my listening to Nadia's Anniversary Album: the playing is uniformly superb—one surprise after another. I never knew she was such a wonderful pianist. How I regret not having heard her before, so I could *tell* her, and pay homage, and spoil her more than I used to. She was something else. I miss her.

Josef Raieff, 1985
The Juilliard School

KABALEVSKY
Twenty-Four Preludes, Op. 38

Nadia Reisenberg, piano.
WESTMINSTER 18095. 12-in. $4.98 (or $3.98).

The Kabalevsky *Preludes*, composed in 1946, are attractive, easily accessible works. Breaking no creative paths, they might have been composed by Mussorgsky, with only a few more modern touches of harmony added, for they have considerable Russian flavor through folkish turns in the melodies and folk-dance rhythms. They follow the pattern of Chopin's Preludes in the matter of key sequence (C, A minor, G, etc.), and vary sharply in mood. Many are technically difficult and make good études, effective for recital purposes and worth pianists' attention. The set finds a wholly persuasive exponent in the expert Nadia Reisenberg.
R. E.

LAS MAZURCAS DE CHOPIN

Consideradas como grupo de obras
bajo la misma denominación, los
estudios y las mazurcas son teni-
dos por los que saben como lo
más grande de su obra. Los pri-
meros han sido harto favorecidos
por las firmas fonográficas, igual
que por los intérpretes, mientras
que las segundas apenas aparecen
ocasionalmente en los programas
con la desesperante falta de indi-
cación del número. Y en discos
solo había la versión magnífica de
Rubinstein, no trasladada al mer-
cado no prohibido. Ahora West-
minster-Fonotón nos entregan la
serie completa a través de las ma-
nos mágicas de Nadia Reisenberg.
El primer disco (18830) contiene
veintidós, el segundo (18831) vein-
te, y la primera cara del tercero
(18832) las catorce restantes, en
una cara, y en la otra la Barca-
rola, la Berceuse y el Allegro de
Concierto en La, opus 46, que es
primicia. Se incluyen en esta serie cinco mazurcas posteriores a
la colección original. Y se disfru-
ta, además de la belleza inmortal
de estas obras en parte apenas
familiares, de una versión admirable.

SUR VOTRE PICK-UP

On ne joue guère, en France,
les sonates pour piano de Haydn
et les discophiles ne se montrent
guère plus empressés à les décou-
vrir que les habitués de nos salles
de concert. La 37• en ré est un
parfait régal. Ecrite, on le sent,
avec plaisir par son auteur, elle
en procure autant qu'elle en reçut
de lui tant son aisance mélodique,
son charme sans apprêt, sa déli-
catesse ont de séduction.

Mme Nadia Reisenberg la joue
bien, avec sobriété, élégance et
pureté. Ce n'est là qu'un tout
petit disque par le format. Raison
de plus pour y goûter. Je serais
bien étonné qu'il ne gagnât au
Josef Haydn des Sonates quelque
amateur sans préjugés qui n'y
avait pas encore autrement (1).

10月新譜

Hi-Fiラボラトリー・シリーズより

☆ムソルグスキー(ラベル編曲)(ML5130)

組曲「展覧会の絵」

アルトゥール・ロジンスキー指揮 ロンドン・フィルハアモニイ交響楽団

☆ヴィオッティ(ML 5132A面)

ヴァイオリン協奏曲 22番 イ短調

★ナルディーニ(ML 5132B面)

ヴァイオリン協奏曲 ホ短調

ペーター・リバール(V) クレメンス・ダヒンデン指揮
ウィンテルトゥール交響楽団

☆ハイドン(ML 5133)

弦楽四重奏曲 作品64—5「ひばり」

弦楽四重奏曲 作品64—2

最高の芸術 最高の録音

Westminster Hi-Fi

Пластинки Нади Рейзенберг

БРАМС — Соната ми-бемоль мажор и Соната фа-минор
Надя Рейзенберг, рояль; Поль Доктор, виолончель
XWN 18114

ШОПЕН — Ноктюрны №№1-11 XWN 18256

ШОПЕН — Ноктюрны №№12-20 XWN 18257

ГАЙДН — Тема с вариациями в фа-миноре и др.
XWN 18057

ГАЙДН — Фортепианные сонаты №№6, 37 и 50
XWN 18357

КАБАЛЕВСКИЙ — 24 прелюда XWN 18095

РАХМАНИНОВ — Колокола, Полька и др. XWN 18209

ЧАЙКОВСКИЙ — 12 фортепианных произведений
XWN 18005

ПЛАСТИНКА НАДИ РЕЙЗЕНБЕРГ И АРТУРА БАЛЬЗАМА

Фирма «Мюзикал Хэритэдж Сосайти», во главе которой стоит д-р М. А. Найда, специализировалась на выпуске долгоиграющих пластинок музыки 17 и 18 веков.

Только что эта фирма выпустила новую пластинку, наигранную двумя известными пианистами Надей Рейзенберг и Артуром Бальзамом — фортепианные сонаты Моцарта для четырех рук. На пластинке зарегистрированы «Соната в Фа мажоре», «Соната Си бемоль мажоре» и «Адажо и аллегро».

WCBS-TV

New York Key Station for the CBS Television Network

485 MADISON AVENUE, NEW YORK 22, N. Y. PLAZA 1-2345

June 18, 1956

Miss Nadia Reisenberg
317 West 83rd Street,
New York, N. Y.

Dear Miss Reisenberg:

Everyone knows what a superb artist you are but it was a pleasure for me as producer to show your lovely and warm personality to a whole new audience. I hope the experience gave you as much pleasure as it gave to the people watching and to me.

Yours sincerely,

Lewis Freedman

Lewis Freedman
Producer, "Camera Three"

LF:ke

THE MANNES COLLEGE OF MUSIC

Tuesday, March 31, 1959—8:30 P.M.

SECOND FACULTY CONCERT
for the benefit of
THE DEVELOPMENT FUND

WILLIAM KROLL, *Violinist*
NADIA REISENBERG, *Pianist*
SHIRLEY VAN BRUNT, *Pianist*
RUDOLPH FELLNER, *Pianist*

MARTIAL SINGHER, *Baritone*
DOROTHY BERGQUIST, *Soprano*
GEORGIANE GREGERSEN, *Mezzo-Soprano*
WAYNE CONNER, *Tenor*

JOHANNES BRAHMS — 1833 - 1897

I

Liebeslieder Waltzes, Opus 52 for Vocal Quartet with Piano Four Hands

II

Vier ernste Gesange (Four Serious Songs), Opus 121
"One thing befalleth the beasts" (Ecclesiastes III: 19-22)
"So I returned" (Ecclesiastes IV: 1-3)
"O death, how bitter" (Ecclesiastes XLI)
"Though I speak with the tongues of men" (Corinthians XIII: 1-3, 12 13)

INTERMISSION

III

Sonata in G major, Opus 78 for Violin and Piano
Vivace ma non troppo
Adagio
Allegro molto moderato

The third Faculty Concert in the Brahms Series will be on Wednesday, April 8th.

LEOPOLD MANNES
120 EAST 75TH STREET
NEW YORK 21, N. Y.

May 4. 1959

Dear Nadia,

Some days have passed since last Wednesday, but my sister and I can never forget the music on that occasion and the very beautiful part which you contributed to it. It was ideally fitting and we were greatly moved. It is hard to put our deep gratitude into words but I hope you can imagine how we feel.

Most sincerely
Leopold

MUNICIPAL CONCERTS ORCHESTRA

JULIUS GROSSMAN
CONDUCTOR

NADIA REISENBERG
PIANIST

Thursday Evening, November 12, 1964 at 8:30

TOWN HALL

123 WEST 43 STREET

BENEFIT CONCERT FOR MUNICIPAL CONCERTS INC.

CRITIC-AT-LARGE by Byron Belt

Mme. Reisenberg Stylish

Elisabeth Schwarzkopf, today's reigning queen of song, was heard last night in recital at Philharmonic Hall. At Saturday's matinee at the Metropolitan Opera one of this era's new breed of operatic glamor gals, soprano Anna Moffo, starred in Donizetti's "Lucia di Lammermoor."

Both lovely ladies must give precedence in this review, however, to one of the great women of the keyboard. Nadia Reisenberg made one of her shamefully rare appearances Saturday night performing the Piano Quartet in B-flat of Charles Camille Saint-Saens with the American String Trio.

We need the mature, gracious artistry of Mme. Reisenberg to help put into perspective today's flashy, generally uninteresting young pianistic virtuosi who play so brilliantly and make so little music. All Nadia Reisenberg does is to play with a warm, stylish and technical assurance that permits music to unfold as naturally as a beautiful blossom.

The Saint-Saens quartet is a sweetly flowing bonbon that was a welcome addition to the gratifying revival of the Frenchman, whose music has been too long out of fashion. Considering that the pianist learned the piece in less than three weeks in honor of the late Arthur Loesser, who was to have been the guest artist, Mme. Reisenberg's accomplishment was even more astonishing. May she be heard again soon, and often!

* * *

Long Island Press, 2/3/69

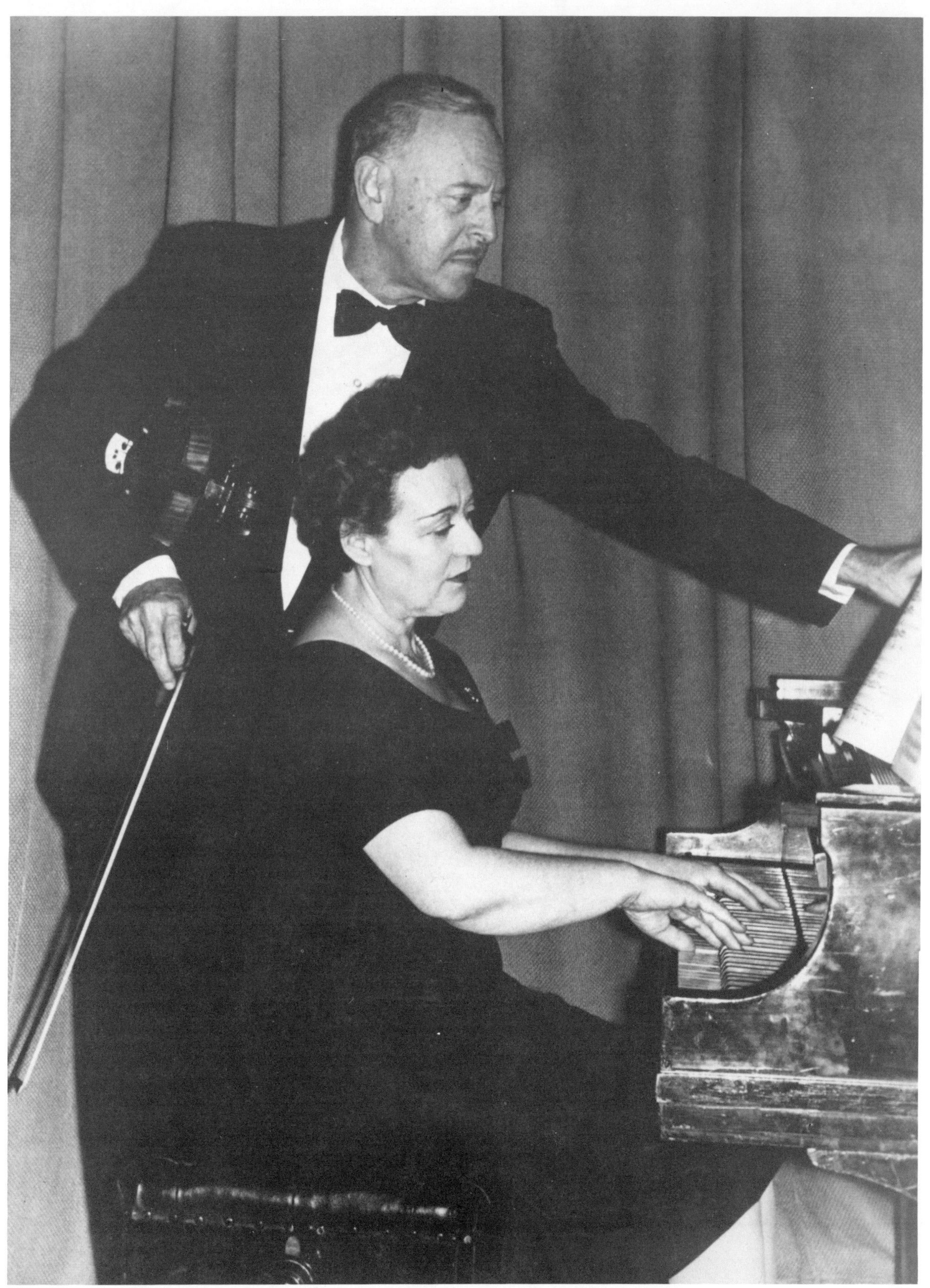

MAVERICK SUNDAY CONCERTS

Maverick Concert Hall Woodstock, New York

SUNDAY, SEPT. 3, 1961 at 3 P.M.

WILLIAM KROLL, Violin
NADIA REISENBERG, Piano

* * *

- P R O G R A M -

Sonata in C Major . . . . Mozart

Allegro vivace
Andante sostenuto
Rondo - allegro

Sonata in D Minor, Opus 108 . . . . . .Brahms

Allegro
Adagio
Un poco presto e con sentimento
Presto agitato

-Intermission-

Sonata Opus 18 . . . . . . .Richard Strauss

Allegro ma non troppo
Improvisation - andante cantabile
Finale - andante allegro

* * *

Admission $1.50 Students $1.00
Block of ten tickets $12.00

* SMOKING ON TERRACE ONLY *

WIGMORE HALL
WIGMORE STREET, W.1

Monday, 16th October 1961 at 7.30 p.m.

A RECITAL BY THE DISTINGUISHED DUO

WILLIAM

KROLL

(violin)

NADIA

REISENBERG

(pianoforte)

CONCERT MANAGEMENT: **WILFRID VAN WYCK LTD.**

TICKETS: Reserved 10/- 7/- Unreserved 4/-

May be obtained at the Box Office, WIGMORE HALL (Weekdays 10-5, Saturdays 10-12.30, WELbeck 2141); CHAPPELL'S Box Office, 50 New Bond Street, W.1. (MAYfair 7600) and usual Agents

London Daily Telegraph, 10/17/61

GIFTED PLAYERS
Subtle Characterisation

A vivid yet subtle characterisation of the music was an outstanding feature of the recital given by William Kroll, violin, and Nadia Reisenberg, piano, at Wigmore Hall last night.

These gifted players from America made it abundantly apparent that they do not impose their own style on the works they perform but find just the right interpretative manner to match the composer's invention.

It was a great pleasure to hear their clear understanding of the very different demands made by the dynamic drama of Beethoven's C minor sonata, Op. 30, No. 2, and the dramatic lyricism of Brahms's D minor sonata, Op. 108.

No less convincing was the appropriate note of jaunty good humour they struck in Prokofiev's sonata, Op. 94, a strong performance of which almost concealed the work's many weaknesses.

The ensemble of this distinguished duo was so closely knit that one hesitates to speak of them as anything but a unity. Now and again the warm, impulsive playing of Miss Reisenberg tended to emphasise the wiry timbre of her partner.

But in the slow movement of the Brahms, Mr. Kroll produced the generous tone that was often absent elsewhere. D. C. P. M.

The Times (London), 10/17/61

A DISTINGUISHED DUO

The adjective distinguished is printed all too freely by managers on their artists' handbills, but in the case of Mr. William Kroll and Miss Nadia Reisenberg, the "distinguished duo" who played at Wigmore Hall last night, it could not have been better deserved - in an autumn season in which violin and piano recitals have been unusually frequent occurrences in this hall, theirs was outstandingly pleasurable.

In the first place, both Mr. Kroll and Miss Reisenberg (who now live and work in America) are individually accomplished and confident as technicians; in the second place they are closely adjusted to each other as a duo—Miss Reisenberg cleverly curbed what now and again seemed like a temptation to wear the trousers.

Last and most important, they have a very direct and strong understanding of the style and musical purpose of each composer they tackle, or so it was last night in sonatas in C minor, Op. 30, by Beethoven, D minor, Op. 108, by Brahms, and D major, Op. 94, by Prokofiev.

Immediately their dramatic dynamic contrasts and rhythmic tension disclosed all that the key of C minor meant to Beethoven, while at the same time their far-sighted phrasing left no doubt of their grasp of each movement's larger architectural span. To Brahms they brought a mellower and more expansive romanticism, and they caught the insouciant charm and gusto of Prokofiev without brashness.

Nadia Reisenberg erhielt ihre Grundausbildung am Kaiserlichen Konservatorium in St. Petersburg und studierte später in den Vereinigten Staaten unter dem Liszt-Schüler Alexander Lambert und dem berühmten Josef Hofmann. Als Solistin spielte sie mit fast allen großen Sinfonie-Orchestern unter Dirigenten wie Barbirolli, Koussevitzky, Rodzinsky u. a. Drei Jahre lehrte sie am Curtis Institute of Music in Philadelphia und ist nun Professor am Mannes College of Music in New York. Nicht nur als Solistin hat sich Nadia Reisenberg einen Namen in den Vereinigten Staaten und Europa geschaffen, sondern auch als Kammermusikerin (Budapest Streich-Quartett u. a.)

William Kroll, in New York City geboren, studierte am Königlichen Konservatorium in Berlin und am Institute of Musical Art, wo er die höchsten Auszeichnungen erhielt und den Loeb-Preis und die Loeb-Medaille gewann. Er spielte im Elshuco Trio und im Coolidge Quartet, bevor er 1945 das Kroll String Quartet gründete. Für seine Verdienste im Rahmen der Kammermusik wurde ihm die Coolidge Medaille der Library of Congress verliehen. Er unterrichtete am Institute of Musical Art, der Universität von Texas und der Universität von Southern California. Zur Zeit lehrt er am Hartt College of Music, am Peabody Conservatory of Music und am Mannes College of Music in New York. Sieben Jahre war er Leiter der Abteilung Kammermusik beim Berkshire Festival in Tanglewood. Als Solist und Kammermusiker bereiste William Kroll die Vereinigten Staaten, Mexiko, Kanada und Europa.

SONNTAG
29. OKTOBER 1961
17 UHR
AMERIKA-HAUS
GROSSER SAAL

DUO-ABEND

WILLIAM KROLL
VIOLINE

NADIA REISENBERG
KLAVIER

Amsterdam, Concertgebouw (kl. z.) – **Den Haag, Diligentia**

Dinsdag 17 oktober 1961 - 20.15 uur — **Donderdag 19 oktober 1961 - 20.00**

Sonaten-avond

WILLIAM **NADIA**

KROLL-REISENBERG

VIOOL **PIANO**

Kaarten à f 1,00 en f 2,00 (inclusief rechten). Abonnees „Mededelingen/Muzenforum" toegang à f 0,50 p. persoon.

Amsterdam
Kaartverkoop en plaatsbespreken vanaf zaterdag 14 oktober 10 uur aan het Concertgebouw. Tel 71.83.45

Steinwayconcertvleugel: vert. Fa. C. C. Bender en Fa. Kettner & Duwaer

NEDERLANDSCHE CONCERTDIRECTIE J. BEEK

Den Haag
Kaartverkoop en plaatsbespreken dagelijks van 10-4 uur aan Diligentia. Tel. 11.76.47

Steinwayconcertvleugel: Fa. C. F. van der Does

Nederlandsche Concertdirectie J. Beek – Den Haag – Tel. 63.37.87-11.63.56

L. v. Beethoven
1770-1827

Sonate No. 7 in c kl.t., op. 30 no. 2
Allegro con brio
Adagio cantabile
Scherzo - Allegro
Finale - Allegro

S. Prokofieff
1891-1953

Sonate in D gr. t., op. 94
Moderato
Scherzo-presto
Andante
Allegro con brio

Pauze

J. Brahms
1833-1897

Sonate No. 3 in d kl. t., op. 108
Allegro
Adagio
Un poco presto e con sentimento
Presto-Agitato

Das Amerika-Haus Kassel erlaubt sich, zu einem

Duo-Abend

mit

William Kroll – Violine

Nadia Reisenberg – Klavier

einzuladen.

Das Konzert findet am Freitag, dem 20. Oktober um 20.00 Uhr im Saal des Amerika-Hauses statt.

Henry W. A. Reinert
Direktor

PROGRAMM

Ludwig van Beethoven – Sonate op. 30 Nr. 2 c-moll
Allegro con brio
Adagio cantabile
Scherzo – Allegro
Finale – Allegro

Serge Prokofieff – Sonate op. 94a D-dur
Moderato
Presto
Andante
Allegro con brio

PAUSE

Norman Dello Joio – Variations and Capriccio

Johannes Brahms – Sonate op. 108 d-moll
Allegro
Adagio
Un poco presto e con sentimento
Presto agitato

KONZERTSAAL DER HOCHSCHULE FÜR MUSIK

Sonnabend, 28. Oktober 1961

2. MEISTERKONZERT

DUO

WILLIAM KROLL

USA (Violine)

NADIA REISENBERG

USA (Klavier)

VORANZEIGE

Konzertsaal der Hochschule für Musik, Sonnabend, 18 November 1961, 20 Uhr

3. Meisterkonzert

QUARTETTO DI ROMA

Mozart, Mendelssohn, Brahms

Karten, soweit vorhanden, Kasse der Hochschule für Musik und Theaterkassen.

amerika-haus · essen

KAMMERMUSIKABEND

mit

William Kroll, Violine

und

Nadia Reisenberg, Klavier

PROGRAMM

Sonate op. 30, No. 2 c-moll . . . . Ludwig van Beethoven
Allegro con brio
Adagio cantabile
Scherzo - Allegro
Finale - Allegro

Sonate op. 94a D-Dur . . . . Serge Prokofieff
Moderato
Presto
Andante
Allegro con brio

PAUSE

Variationen und Capriccio . . . . Norman Dello Joio

Sonate op. 108 d-moll . . . . Johannes Brahms
Allegro
Adagio
Un poco presto e consentimento
Presto agitato

FIFTEENTH CONCERT SERIES 1968-69

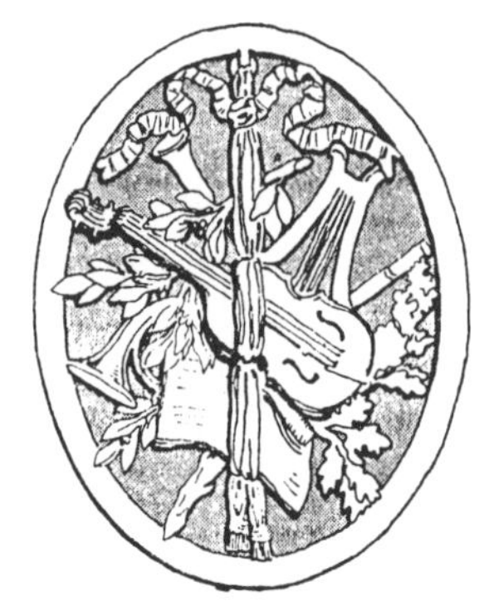

Friends of Music

presents

Nadia Reisenberg
and
William Kroll

Piano and Violin

Saturday November 9, 1968 at 8:30 p.m.

BRIARCLIFF HIGH SCHOOL- BRIARCLIFF MANOR, N.Y.

Nadia enjoyed playing with "Fritz" enormously. She respected him greatly as a chamber music partner, and had the most positive feelings toward him as a colleague.

Clara, 1985

Clara, William Kroll, and NR

Bob, NR, and William Kroll at Clara's home, New Year's Eve, 1968

Page 38, Wednesday, August 5, 1970

WILLIAM KROLL NADIA REISENBERG AT MAVERICK SUNDAY

BASIL ELIESCU

The Sunday August 9 concert will celebrate the annual return to the sylvan Maverick Hall, of one of Woodstock's most beloved musicians William Kroll. This time in the company of the celebrated pianist Nadia Reisenberg. Favorite performer here Mr. Kroll, with Mme Reisenberg, will present works by Beethoven, Richard Strauss, and Cesar Frank. The Strauss, though presented here some years ago, is novelty as Strauss is better known for his symphonic and operatic works. It should be very interesting to listen to the interpretation the Kroll-Reisenberg team will give this rarity. As we remember thelast time Mr. Kroll played it was most definitely a great pleasure to hear it. It was gorgeous, and he played the long soaring lines with utter ease and conviction. He was absolutely at home in the rambling structures and bigger-than life gestures of this powerful and passionate work.

Of Na Dia Reisenberg, not much has to be said. She is well known world-wide for her concert stage appearances. Having received her initial education at the Imperial Conservatory in St. Petersburg, she continued studying with the famous Joseph Hofman. She has appeared as soloist with all the great symphony orchestras under the direction of Barbirolli, Kousevitzky, Rodzinsky. She has taught at the Curtis Institute in Philadelphia, and now she is professor at the Mannes College of Music in New York. Mme Reisenberg is widely known in Europe and the United States not only as a solo pianist, but also as a chamber music player appearing often with the Budapest String Quartet, as well as other groups. She has recorded extensively, including Russian works of Tschaikowsky, Moussrgsky, Rachmaninoff as well as all the Nocturnes and Mazurkas of Chopin, and the Haydn Sonatas.

William M. Kroll, "[illegible] and friends, always arouses an atmosphere

William Kroll and NR, Woodstock, N.Y.

Nadia Reisenberg - long may you reign -
Queen of music, both fancy and pleign!
Among days I call great
Is your Maverick date - -
May you come back ageign and ageign!

You and stagemate today, William Kroll,
Are performers of such skill and soll
That I'm sure as a pair
You'll make music as rair
As the nectar in Ganymede's boll.

John C. Bagley
John C. Bagley
The Heterographicalimericist

Catskill, N.Y.
August 15, 1971.

Family Interlude V

NR with Steve, second grandson, born June 6, 1959

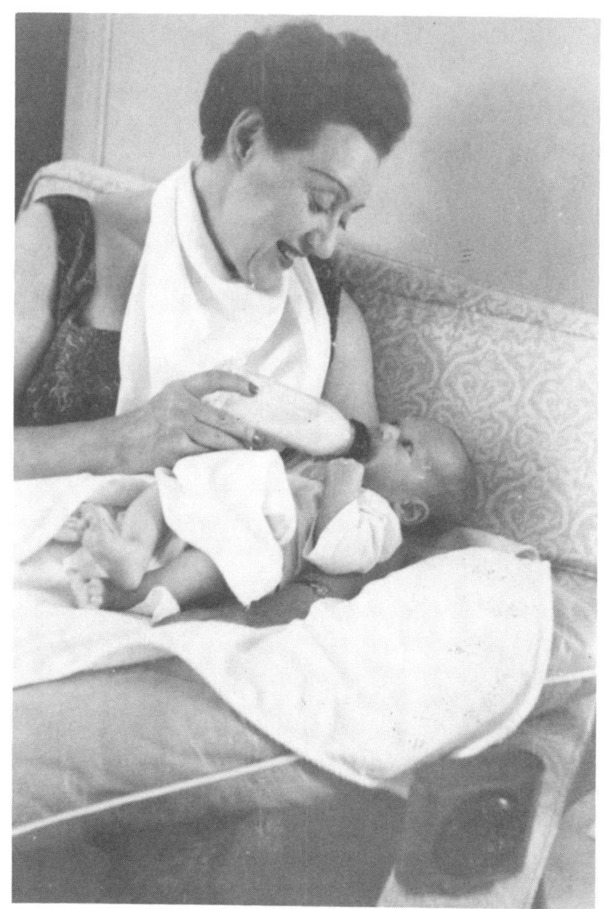

NR with Steve (right) and third grandson, Peter, born March 25, 1964 (left)

Plants there better be in my apartment, and flowers too, and I have a little terrace which is my pride...

NR, 1980

NR with Mstislav Rostropovich

Ruth Sherman, NR, and Newta

NR, grandson Peter, and Zalman (Ziama) Gershuni

With grandsons David, Peter, and Steve

Erna and Walter Fisher (Ruth's parents), Ruth, Alex, David, and NR

NR and David Sherman

Nadia Prujan and NR, 1964

Standing: Andrei Sedych, Jennie Grey (Zwibak), and NR; seated: Fanny Racine and Katya Litton

NR between the two mothers-in-law, Erna Fisher and Bronia Gershuni

Before the Revolution in Russia, the orthodox religion was dominant, and during Lent, many kinds of music, theatre, dance and other entertainments weren't permitted. Since artists had nothing to do in this period, they began putting on private shows, just for fun, to entertain each other. An evening like this was called a Kapustnik, after "Kapusta," the Russian word for cabbage,and therefore the most boring ordinary thing because this was the main dish during lent—cabbage in borscht, cabbage as a vegetable, cabbage with rice, cabbage with anything you can imagine. This kind of clowning-around evening had never been done among Russian emigrants in America, but we had a lot of wonderful talent here in the 1950s, so we arranged our own Kapustnik.

Andrei Sedych, 1985

NR and William Kroll

NR, Andrei Sedych, and Dr. Weinberg

One time I pretended to be a pianist. Nadia, dressed as a little girl, was the page-turner who finally took over at the keyboard, and Dr. Weinberg, who was well known in the Russian community at that time, came out with some gypsy songs on the violin. We were young, and life was pleasant...

Andrei Sedych, 1985

NR, Mara Rosenthal, (?), and Lydia Chaliapin

Master Teacher

Teaching contributes enormously to one's development as a musician. To be able to analyze what is troubling your students (even if it has never been your own problem), to innovate ways of building their potential without stifling their personalities, is very exciting. I find teaching gratifying, interesting, rewarding, and I have never felt in any way that it is somehow a substitution for something which I no longer do. I always loved teaching, even as a youngster in the conservatory. The thing is not simply to do something, but to know *why* you're doing it, why it should be done. I think I've been successful, in the sense that most of my students really are educated and intelligent musicians.

NR, 1974

A good teacher has to have, first of all, the talent for teaching. There are a number of excellent instrumentalists who don't have this talent of helping those who have problems, which perhaps were never their own problems. Then, one of the most important things in developing a young student is knowing how to give them the teaching material: how to understand, let us say, a 14-year-old who may have a magnificent pianistic talent but is still inexperienced, musically speaking; how to satisfy his need to show off what he can do on the piano, and yet make him learn the proper repertoire, show him what stylistic differences are, help him develop good taste, and so on…

NR, 1975

CARNEGIE RECITAL HALL

157 West 57th Street
New York

NADIA REISENBERG

presents

four artist-pupils
in
four piano recitals

on Friday Evenings at 8:30

May 2 ROGER KAMIEN

May 9 LYNNE KLEINBERGER

May 23 MOREY RITT

May 30 EUGENIA HYMAN

TICKETS for single recital
Orchestra (Center and Left Side) $1.80 incl. tax
Orchestra (Right Side) . . . . 1.20 " "
Balcony . . . . .90 " "
TICKETS (for all four recitals)
Orchestra (Center) . . . . . . 4.20 " "

The early teacher is a deciding factor. If the bedrock is not carefully constructed, the pupil is never ready for the "artist teacher," for he cannot understand and evaluate the work the latter gives him. Correct timing begins at the first lesson; the groundwork is set in the first years, and that this be solidly and logically built is more than important. It is indispensable.

NR, 1941

New York Times, 5/3/52

Roger Kamien, Pianist

The first of four recitals by artist-pupils of Nadia Reisenberg was heard at Carnegie Recital Hall last night. Roger Kamian, eighteen, was the first of the young recitalists, and a talented pianist he is.

The program was designed to show various aspects of technique and talent; the Beethoven C minor Variations and the Mozart Sonata in D major demonstrated a clear, crisp technical proficiency that would have done credit to any pianist. The touch and tone in the Mozart were elegant indeed; he Beethoven, dynamically speaking, was an exceptionally lucid interpretation.

Performance and interpretation of the classic repertory such as these Mozart and Beethoven works is so standardized that it is difficult, when young players are as well taught as this one is, to decide where the teacher leaves off and the pupil begins. But in a modern romantic piece like the Kabalevski Sonata Op. 46, where no standardization in concept has taken place, one can readily estimate the extent of the performer's judgment, sensitivity and musical insight. It was here that one perceived that Mr. Kamien is a musical individual and one from which much can be expected.

P. G.-H.

Nadia has tremendous pianistic gifts, obviously, but for me she was primarily a musician of great depth. She was always examining what the composer was saying; she made me—as a violist—begin to examine much more thoroughly what was behind the printed page. This is what makes her a marvelous chamber-music player, and a teacher without peer.

Milton Katims, 1980

It's very important to let talent ripen. The main trouble with being very competitive is that you usually don't have the time to amass the necessary repertoire. So often pianists win competitions very early, before having had a very thorough education, and when they are "lucky" enough to have many concerts, what happens is they will have to play the same three concertos instead of taking the time to develop. In my time, actually, there were no such things as competitions: you became an artist when you were ripe enough to become an artist, and you started your career that way, and it was solid. I feel it is extremely important to nurture a young talent, to develop it with love, with patience and then, only when the time comes, start the career.

NR, 1975

NR at home, with pupils, 1949

Top Row (left to right): (?), Joseph Liebling, (?), (?) Anita Notarius, (?), Eugenia Hyman, Esther Fernandez, Gilman Collier, Lynn Kleinberger, Ida Hartmann DeMotte, Roger Kamien. Seated: Phyllis Rappeport, (?), NR, Marlene Allen, Jeanne Singer. Front Row: Bob.

My deep love of music is mainly because of Nadia. I have had her photograph on my piano for about thirty-five years, and when I play or teach, I think of her, her goodness, her spirit, her loving of living...
Lynn Kleinberger (Rosen), (top row, third from right), 1983

I shall always regard Nadia as one of the prime formative influences in my musical life. This went far beyond mere piano instruction and extended deeply into the most important aspect of my career, composing. For not only did she leave me with a piano technique so thoroughly grounded that I can call upon it whenever needed, but her wonderfully musical and always constructive criticisms spilled over into several of my earliest compositions. Never content with confining herself merely to technical matters, she gave me in the course of these lessons some of the finest composing advice I ever received. She was able to do this because of her magnificent, but totally objective, musicianship. Her remarkable gifts, so unstintingly bestowed upon her pupils, will be a predominant part of our musical lives throughout all the length of our days.
Gilman Collier (top row, fourth from right), 1983

Throughout the years, I realize that I have been playing for her whenever I am at the piano. She was not only my teacher, but my friend...
Muriel Elliott , 1983

She had an effect and influence over me that proved to be the strongest I have ever experienced: she taught me to think, to listen, and finally to be totally honest in my self-evaluation. There is not a phrase that I play that I don't ask myself, "What would she say about this? What would she tell me to listen for? Would she consider this my best? No, I can do it better..." She was a uniquely gifted pianist and an absolutely genuine person. I'm sorry I never told her that I loved her.
Anna Manicone , 1983

Our beloved Nadia, a woman of valor, great artistry, and poetic sensitivity. We will always try to be the kind of person she was—so uniquely endowed with a humanity that lights up our paths. The song of her music will go on in our hearts forever...
Bella Shumiatcher 1983

We all knew her as a consummate pianist and a musician's musician. Those of us who were fortunate enough to study with her also knew her as an inspiring teacher, an all-around person with many interests in addition to music, and a good friend. Nadia was the most influential person in my life. I'll remember her always with love, respect, and gratitude.
Gershen Konikow, 1983

Tulsa Daily World, 6/6/52

Enthusiasm High in Seminar Classes of Nadia Reisenberg

By MAURINE HALLIBURTON

The delight of any teacher is an enthusiastic pupil, and Nadia Reisenberg, concert pianist and teacher, can find no fault on that score with the teachers and pianists who are taking classes under her. These "pupils" come early, stay late and are eager to perform, to be instructed and even to be corrected by the eminent musician, in the seminar series sponsored by the Piano Study club.

Miss Reisenberg's enthusiasm matches their own. In an interview Thursday at B'nai Emunah synagogue, where the seminar is in progress, she said that she is delighted with the quality of musicianship shown by those taking the classes. She likes, too, the eagerness, warmth, cooperativeness and sense of humor that are shown during the informal sessions. In such atmosphere she does her best work, she says, and gets the best work from others.

It is the atmosphere she naturally evokes, as she, too, is eager and warm and cooperative. She spares no one in criticising what is wrong, but the criticism is given with friendly interest and because it is equally as valuable as the earned praise for good work. And she scolds, sometimes, when she finds that her "pupils" are neglecting their practising, a fault that teachers develop as work piles up and time grows short.

They are warned, "Whatever else you give up, you must take an hour or an hour and a half at least each day for practice."

No one objects to this gentle scolding as all know it is of the utmost importance to any professional or amateur musician, and their taking the seminar in the midst of their own busy schedules proves their seriousness of purpose.

Since her debut in 1923, Miss Reisenberg has been recognized as one of the foremost women pianists. She has appeared with outstanding symphony orchestras and has concertized in the great cities here and abroad. She was on the faculty of the Curtis Institute of Music three years and since then has been teaching in New York City, turning out artists who have gained recognition in the concert field and as teachers on the faculties of colleges and universities or in their own studios.

From Tulsa she will go to Los Angeles, where she will teach in the University of California and Los Angeles. She will return then to New York for a series of broadcasts before taking up her winter schedule of teaching. Her students are both teachers and performers.

The Tulsa seminar series will continue through June 14, with classes from 9 a.m. until 1 p.m. daily, besides work with those taking private lessons. Among those in the class Thursday were Jewel Major Roche and Mr. and Mrs. Herbert Ricker, teachers from Oklahoma City. Mrs. George Baker comes from Talala for daily work and Mrs. Wesley Rose comes every other day from Hominy.

Besides the teaching, Miss Reisenberg finds herself in demand for many social events in music circles. A reception in her honor given Thursday night by the Sigma Alpha Iota musical fraternity was held in the home of Mrs. B. W. Griffith. Miss Reisenberg is a national honorary member of the organization. Mrs. A. G. Oliphant entertained for the artist at a dinner Wednesday night at Southern Hills country club, and Miss Gertrude Clarke was hostess to a small group at dinner Tuesday night. Mrs. John Leavell will entertain in her honor at dinner Saturday night and Mrs. B. W. Griffith at luncheon the same day.

Enthusiasm for Tulsa and surprise that it is so "green," was expressed by the visitor. She said she had expected a bleak, treeless country here. Also she had been advised against bringing a raincoat to Oklahoma, but did so, anyway, and has found it a very necessary article of dress since her arrival Monday.

"Pan Pipes" of Sigma Alpha Iota, November 1952

With Our Honorary Artists

New York pianist **Nadia Reisenberg** was in Tulsa, Oklahoma for Master Classes two weeks in June, brought to Tulsa by the Piano Study Club, SAI Rosalie Talbott, president. During her stay SAIs had open house for her at the home of Mrs. B. W. Griffeth, patroness. Following her two weeks at Tulsa, Miss Reisenberg spent the entire summer teaching at the University of Southern California. Among appearances while in Los Angeles were two concerts with Joseph Schuster, 'cellist, playing all of Beethoven's works for piano and 'cello; with the Bovard String Quartet in Schumann's *Quintet for Piano and Strings*, Opus 44; with clarinetist Mitchell Lurie and harpist Catherine Johnk in a concert of chamber music in the Los Angeles County Museum; and at the 42nd annual convention of the Music Teachers Association of California held at Riverside June 30 to July 3.

You must be personally interested in your pupils — know something about them as people as well as their musical ability. I must know what my students are thinking — some are like sponges and can absorb information ad infinitum; others can execute very well a certain suggestion, but not many suggestions at once. Some need discipline, others need encouragement...

NR, 1956

You have to train rounded musicians. I keep a book where each student's repertoir is right there at my fingertips, and I see to it that there is a balance, just like a balanced diet; That beginning with baroque, through the classical and romantic periods, and on to impressionistic and contemporary, they have an equally well-developed representation of musical styles.

NR, 1981

I never just play for students. First comes an explanation, then an illustration of the explanation, which very often involves a physical motion that would be clearer if they can see it. You don't ever play in order for them to imitate. I hope, I *hope* that every one of my many students plays his own way, shows his own personality.

NR, 1975

Advice to Teachers:

Rather do without breakfast, or take one student less, but for heaven's sake, don't ever leave the piano, even if you only practice an hour a day. First of all, you have to grow, musically as well as technically. A teacher, like a performer, either develops or slides backwards. We cannot freeze our degree of excellence — we either go up (and I hope I am going up every day of my life) or we go down. We don't stay put . . .

NR, 1981

Miss Reisenberg helped me tremendously. I've learned everything from her...

Myung Whun Chung, 1976

Shulamit and NR, 1966

NR and Hai-Kyung Suh at Mannes School, 1977

NR (left) and Morey Ritt (right) played the Rachmaninoff Suite No. 2 at an evening honoring Andrei Sedych, Carnegie Recital Hall, March 14, 1962

All of Nadia's students shared her special gift to us. During lessons, she would present, effortlessly, perfect demonstrations of virtually any repertoire we brought in. This playing was our inspiration and a continuing source of wonder, particularly when the music was unfamiliar, a new work that she could never have seen or heard before. While most of us have memories of favorite performances by Nadia, and while there are, fortunately, many beautiful recordings, we former students remember the playing that was just for us. This portion of Nadia's musical legacy is ours alone.

Morey Ritt, 1983

Juliana Osinchuk and NR at the Juilliard School

Craig Burrell, Chairman of the Board, Mannes College of Music; Rise Stevens, President of Mannes; Alice Tully; and NR

The Mannes College of Music

Risë Stevens, *President*

Commencement

Monday, June 6, 1977, at 3 p. m.

157 East 74th Street
New York City

PROGRAM

Incidental Music by the Mannes Brass Ensemble
Simon Karasick, *Musical Director*

I
Introductory Remarks
Craig D. Burrell. M. D., *Chairman of the Board*

II
Presentation of Awards and Honors
David Tcimpidis, *Dean*

III
Sonata for Oboe, *Munter* — Paul Hindemith (1895-1963)
David Woolsey, *Oboe*
Paul Suits, *Piano*

Ondine (from Gaspard de la Nuit) — Maurice Ravel (1875-1937)
Jeanette Kim, *Piano*

IV
Presentation of Honorary Degree
Miss Alice Tully, *Doctor of Music*

V
Commencement Address
Mme. Nadia Reisenberg

VI
Presentation of Diplomas

I have nothing but the highest praise for that wonderful woman. Nadia Reisenberg was such a fabulous artist, and a staunch advisor to me at the Mannes College during my years there as president. When I asked Vladimir Horowitz to join our faculty, he chose Mme. Reisenberg to make the final choice of the pianist she felt should work with him. God rest her soul, she was the greatest of the great.

Risë Stevens, 1985

Performance is very important for students. Ours is a performing art, and we have to get used to getting up on the stage, even how to bow; playing is part of the study. I very often call my whole class together before exams and let them play for each other, and they criticize each other. I find this is very important.

NR, 1981

The Class of 1981-82. A party in December 1981. Top row, left to right: Dal-Ok Han, Chinho Kim,Gwan-Ying Wu, Karina Eberl, Robert Bank, Lev Ostrovsky, Alan Moverman, Tai-Chang Chen, Elena Leonova. Middle row: Elena Klionsky, Eun-Soo Son, Hae-Young Chun, NR, Boris Slutsky. Bottom row: Konstantin Orbelian, Vladimir Zaitsev, Dmitry Rakhmanov.

I will always be very proud that I'm her student, and my students are proud to be "grand-students" of Nadia Reisenberg. She was a dear friend to me, like a close member of my family. I love her and I remember her. In music, I will try to do every possible thing so that she would approve of it.

Elena Leonova (top right), 1983

My ten years of study and friendship have been the most meaningful, the most encouraging, and most of all the most totally musical experiences of my life. In these days of such commercialism and trivialities in music, Nadia Reisenberg's integrity, poetry, and idealism are my most treasured examples. I will try with all my forces to play as Nadia would have wanted me to...

Vinia Tsopelas, 1983

She taught with such dedication and passion that one felt one's own importance to her. I am happy and thankful to say that Nadia Reisenberg has been the person most sustaining, artistically influential, and reassuring in my adult life. She would not let you go until you had achieved that special musical quality she knew was the essence of the piece. She was truly my dear teacher and my dear friend.

Trudi Super, 1983

My association with her is unmatched by any other in my life. I have been enriched immeasurably, not only through her art, but by her singular qualities as a human being. Miss Reisenberg once told me of the strong mutual love that existed between herself and all of her students, and this is certainly true in my case. She will always be a part of me, and a continuing influence in my life.

Benjamin Bradham, 1983

Madame Reisenberg was my teacher for the past five years and my mentor for as long as I can remember. Her patience, her understanding, and her kindness never failed to impress me. She encouraged me and helped me through so many obstacles I can't even begin to describe them. I am sure she cared about every one of her students, but the faith she had in me kept me going. I was fortunate to have known her and to have learned what I could from this great woman who was my teacher.

Hae Young Chun (middle row third from left), 1983

Every time I would go to my lesson, I would return home with "gold" in my pocket, and I would feel stronger and happier about life. You see, she was not only a great pianist, she was also a warm and generous person...

Mélisande Chauveau, 1983

NR and Dalit Paz Warshaw, 1982

Four Fun Pieces

Composed by: Dalit Paz Warshaw

Fun Going To Miss Reisenberg

We are driving to Manhattan to visit Miss Reisenberg. There is traffic. The horns are honking.

I'm riding up the elevator.

Then I tiptoe to her apartment door. I ring the bell. Miss Reisenberg opens the door. She smiles. She offers me food:

"Do you want cake, and we have hot chocolate, or I'll give you cookies. I'll give you ice cream, we have coffee ice cream, and vanilla, and does your Mommy want something?"

She's very generous.

I play for her my composition, "The Waltz".

Miss Reisenberg shows me her pretty flower garden.

All of a sudden, we hear Daddy's horn honking downstairs. I am very sad.

I have to go. Good Bye, Miss Reisenberg.

Not only do my students get cookies, they choose them! They say "I like this sort, where are they?" or "I like the ones with sesame and for a long time I haven't seen them." Well, why not? Didn't I immediately ask you if you wanted tea or coffee? If I have a guest in my house, I always offer something—it's such a natural thing. Students are probably a little less tense in a home atmosphere, so let there be cookies around . . .
NR, interview with June Le Bell, 1980

Jerusalem Post, 8/3/60

Piano Virtuoso Incognito

Seldom has an outstanding professional personality arrived in our country so unheralded and unknown as the eminent pianist and teacher, Nadia Reisenberg, but rarely has a name spread so fast amongst professional circles.

Nadia Reisenberg came to Israel simply to visit her sister in Haifa. Having fallen in love with Israel at first sight, she remarked that she would like to contribute something to the progress of its musical life.

The first to take advantage of this singular opportunity was the Rubin Academy of Music in Jerusalem, which hastily gathered together anyone available (as the study year had already closed). In five days Miss Reisenberg worked with a group of teachers and students, explaining, analyzing and demonstrating some great works of the piano repertoire, infecting everybody with her enthuasiasm for her vocation and exciting all present with the impact of her personality, the soundness of her musicianship and her pianistic perfection.

From Jerusalem she moved to Zikhron Ya'acov to take a hand in the Fourth Chamber Music Seminar, initiated by the Government Tourist Corporation in close cooperation with the America-Israel Cultural Foundation and the Ministry of Education and Culture. Here she continued with her stimulating teaching, climaxing her stay with a sonata recital together with viola-player Oedoen Partos (see Musical Diary-Page 4).

Miss Reisenberg received her initial education at the Imperial Conservatoire in St. Petersburg, continuing later in the United States with Liszt's pupil Alexander Lambert and the famous Josef Hofmann.

Acclaimed as the most promising of the young pianists after her debut at the Aeolian Hall in New York, she quickly made her way, appearing as soloist with practically all the great symphony orchestras and playing under such as giants Damrosch and Koussevitzky. Specializing in interpreting Russian concertos hitherto unknown in the Western Hemisphere, she did not neglect other contemporaries and astonished everybody with her brilliant performance of all 27 piano concertos by Mozart, which had never been done before and has not been repeated since.

After 3 years of teaching at the Curtis Institute of Music, Philadelphia, she taught privately, and since 1955 has been professor at the Mannes College of Music, New York. Miss Reisenberg is widely known in Europe and the U.S. as an outstanding pianist. Unusual for a solo pianist, she is also a very sought after chamber music player and his often teamed up with the Budapest String Quartet and other instrumentalists.

She has made some 16 LP records (Westminster), amongst them: all the Chopin Mazurkas and Nocturnes; many Haydn Sonatas; works by Tschaikovsky, Moussorgsky, Kabalevsky, and Rachmaninoff.

YOHANAN BOEH

MUSICAL DIARY

Zikhron Ya'acov Concert

Sonata Recital by Oedoen Partos, viola and Nadia Reisenberg, piano (Beit Daniel, Zikhron Ya'acov, July 30). J. S. Bach: Sonata No. 3 in G minor for viola da gamba; Brahms: Sonata in F minor, Opus 129, No. 1; Schubert: Sonata for Arpeggione; Brahms: Sonata in E flat, Opus 120, No. 2.

A CHANCE meeting during the Fourth Chamber Music Seminar at Zikhron Ya'acov resulted in this improvised concert in the charming little recital hall, Beit Lilian, on Saturday night. Oedoen Partos and Nadia Reisenberg teamed up without problems in interpretation, and the two wonderful artists managed in only one rehearsal to prepare a demanding programme and to execute it in an intimate atmosphere as befits best the presentation of chamber music. Bach's music came over in grand manner: the two Brahms Sonatas, especially the second one, were given a thoroughly enjoyable performance, and in the Schubert Sonata Mr. Partos added a few reckless accents as if affected by the general enthusiasm.

The Fourth Summer Seminar for Chamber Music at "Beit Daniel" in Zikhron Ya'acov closes on Friday after three weeks of day-long activities.

This year 40 young students from all over the country and two guests from Cyprus took part in the seminar, planned by the Government Tourist Corporation, the America-Israel Cultural Foundation and the Ministry of Education and Culture. Main instructors were Oedoen Partos and Frank Pelleg, with Yosef Tal, Gary Bertini and the Duo Pianists Eden-Tamir contributing guest lectures and demonstration lessons. The outstanding feature of this year's seminar was the attendance of Nadia Reisenberg, who evoked interest amongst the students.

It is planned to enlarge the number of students next year to 60 by inviting participants from Mediterranean countries and from Africa.

Y. B.

My work with Israeli students, here and in Israel, was a source of great joy to me always. These youngsters are very close to my heart.

NR, 1974

רסיטל לכנור ופסנתר

VIOLIN & PIANO RECITAL

יאיר קלס - כנור

YAIR KLESS - VIOLIN

נדיה רייזנברג - פסנתר

NADIA REINSENBERG - PIANO

Programme — התכנית

BEETHOVEN - — בטהובן -
Sonata No. 7 — סונטה מס׳ 7
in C minor — בדו מינור

SCHUBERT - — שוברט -
Grand Duo in A Major — דואו בלה מז׳ור
Op. 162 — אופ. 162
Allegro moderato - Scherzo - Andantino - Allegro vivace

FRANK - — פרנק -
Sonata in A Major — סונטה בלה מזור
Allegretto ben moderato - Allegro - Recitativo - Fantasia - Allegretto poco mosso

Ebullient pianist, shy fiddler

Sonata Recital by Nadia Reisenberg, piano, and Yair Kless, violin (The Jerusalem Khan—July 23). Beethoven: Sonata in C minor, op. 30, no. 2; Schubert: Fantasy in C major, op. 159; Franck: Sonata in A major.

A BY-PRODUCT of the Rubin Academy Summer Courses, this sonata recital brought together two artists who, in the normal run of things, would probably not perform together because of differences of temperament and approach. Nadia Reisenberg, bubbling over with energy and musical drive, tried to give the music life without openly dominating or drowning out the violinist. Yair Kless, restrained and cool, played his part without any personal expression or emotional participation. In this respect, the pair reminded one slightly of Hephtziba and Yehudi Menuhin in their sonata recitals.

In the Beethoven sonata, the weight is on the piano anyway, so that the combination worked out quite well and, under Nadia Reisenberg's handling, the work was given an appropriate reading. The Schubert — a rather drawn-out work with some weak sections — revealed the inequality of the two artists even more strongly, so that I did not feel inclined to sit through the Cesar Franck sonata, which one hears so often, even in the version with cello. The tail-end of the season and a very hot summer night demand lighter fare.

Nadia Reisenberg is a pianist of high quality, with fluent technique and fine musicality, and a solo recital or pairing with a congenial instrumentalist would have brought out her artistry even more. Yair Kless has his virtues in certain spheres — he is undoubtedly an excellent teacher and may be a good chamber music player—but he seems to be too shy and introverted for solo performances.

YOHANAN BOEHM

I vividly remember hearing Nadia play for the first time: It was in Jerusalem and she performed the Franck sonata with someone whose name I can't recall. Her playing was warm and radiant, direct, spacious, clean, gracious, and strong—the violinist's was somewhat less so (so much the stronger *she*!) As a colleague, she was really a pro—always prompt, always interested, always wearing a good head. She was a delight...

William Masselos, 1983

Nadav,

You seem to think that you are needed in the US, but you should know that here in Israel we need you much more. Come back quickly and for a longer period.

Sincerely
Yours

Yohanan Bader

30/8/62 Jerusalem

Jennie Tourel and NR

Jerusalem Post, 8/10/62

TWO eminent artists from the U.S. — both ladies, both Russian-born, both at the height of their careers — are now in Jerusalem for some six weeks at the invitation of the Rubin Academy, and their open lessons draw students from all over Israel. Nadia Reisenberg teaches piano in the morning, and Jennie Tourel singing in the afternoon. An atmosphere of concentrated attention pervades these classes, which are quite an experience. Nadia Reisenberg makes immediate contact with her students, and her penetrating and lucid remarks and corrections — immediately made on the second piano — draw forth gasps of admiration and unrestrained approval. Stressing neglected inner voices or bass lines, dynamic shades and instrumental colours, she imparts sudden clarity to muddled and unconvincing presentations. Her teaching, seasoned with wit, is never condescending or academic and she spills over into musicology, psychology and proper posture.

קורסים של קיץ באקדמיה למוסיקה בירושלים

מאת מנשה רבינא

היו זמנים, שבהם ידע כל מורה למוסיקה, כי למטרת ההשתלמות עליו לצאת לחוץ לארץ. המורים נאנחו קשות כשהגיע התור לארוז את המזוודות, כי הנסיעה המייגעת הפחידה אותם, אבל מה לא יעשה מורה מסור למען חניכיו, המצפים ממנו הכוונה בטוחה אל פסגות הפארנאסוס? נאנחו ונסעו, ביטלו זמנם מחופשה והקדישו עצמם ללימודים מעמיקים.

הנהלת האקדמיה למוסיקה בירושלים החליטה לפתור בעיית ההשתלמות של המורים למוסיקה בצורה הנוחה וההגיונית ביותר. רוצים להשתלם? — בבקשה. נסדר בשבילכם קורסים של קיץ. לא רוצים לנסוע לחוץ לארץ? — בבקשה, נביא את „חוץ לארץ" אליכם.

וכך הגיעו אלינו מחוץ לארץ שתי מורות מפורסמות, ששמן הולך לפניהן. לעת עתה רק שתיים. להוראה בפסנתר ובזימרה. ההתעניינות הרבה בשני הקורסים האלה תעודד בלי ספק את הנהלת האקדמיה למוסיקה בירושלים להרחיב את המסגרת בעונות הבאות עלינו לטובה. כיום יושבים בירושלים למעלה מ־100 מורים/ות, השותים בצמאון דברי המורות הידועות: הגב' נדיה רייזנברג והגב' ג'ני טורל.

מכל הארץ באו אל הקורסים האלה. מובן מאליו, שרוב המאזינים הם מירושלים הבירה, אלה אינם צריכים לטרוד בנסיעה כל שהיא, וכו. והתורה מוגשת להם בביתם שלהם. מבין האורחים בולטים אנשי חיפה צמאי התורה. באו גם מהגליל ומהעמק. ובפה מלא ניתן לומר, שכל אלה שנרשמו לקורסים באו ללמוד, להאזין, שלא כבקורסים בחוץ לארץ, שבהם לעתים קרובות אין לראות את כל הנרשמים בשעת השעורים, כי בחוץ לארץ יש כוחות מושכים מחוץ לאולם ההוראה. כאן, בכל שעור יושבים כל המאזינים. ובסיפוק רב יש לציין את העובדה, שרבים מאלה שנרשמו לקורס של הגב' נדיה רייזנברג לפסנתר, ביקרו אחרי הצהריים גם בשעורים לזימרה של הגב' ג'ני טורל. ובטוחני, שיותר לא לבקורתנים, אין לי אפשרות לדון על הנעשה שם. בשעורים שבהם הייתי נוכח, הגיעו לאזני קומפוזיציות רומאנטיות בלבד. התפלאתי על כך, אבל נוכחתי לדעת, שמומנט הזימריות של הנגינה בפסנתר ניתן להסברה מעמיקה על יסוד קומפוזיציות אלה.

הגב' נדיה רייזנברג חוננה בכשרון רב להשמיע בפסנתר שלה את השלילי שבנגינת התלמיד, ועל ידי כך להדגים לפניו את הבלתי־רצוי, כשמיד אחריו בא הרצוי בביצועה של המורה. הדגמה זו נעשית בבהירות כזאת, שהתלמיד תופס מיד את הטעון תיקון. שאלה היא, אם יוכל מיד לתקן את המעוות; אבל המורה רייזנברג אינה מחכה לתוצאות מיידיות. כוונתה העיקרית היא להניע את החניך, ועל ידו את כל הנוכחים באולם — מורים ותלמידים גם יחד — למחשבה, לבקורת, לעיון. לדעתה אין להפריד בין ניתוח החומר הטכני ובין ההגשמה האמוציונאלית רוחנית.

מלה אחת לא נזכרה בכל השעורים, שבהם ביקרתי. ודבר זה יכול לעורר תמהון: אף פעם לא דובר על „מגוע" (טושה). נדמה לי, שהמורה רייזנברג רואה מושג זה במזיגה של כל התכונות החיוביות בנגינה של הפסנתר. אם החניך יגיע לשימוש נכון בדוושה (ועל כך מקפידה היא מאוד), אם הוא ידע לגוון נגינתו בדינאמיקה (= כוח הצליל, במקרה זה) הנכונה, אם הוא „ינשום" נכונה ויבין את הפיסוק המיוחד לכל סגנון, אם הוא לא יזלזל בצד הטכני ויקפיד על ביצוע כל צליל בהתאם למשקלו בקו או בקישוט, אם הוא... כי אז ישיג את המגוע הרייזנברגי, המפתיע במידה כזאת את כל הנוכחים באולם.

הגב' ג'ני טורל טוענת בשעורים שלה, שאם התלמיד יהדור לרוח המלים, אם הוא יקדיש תשומת לבו לא רק לביצוע הצלילים הנכונים, כי אם יחפש הבעה ה„ליד" בהתאם לנרמז במלים ובהתאם לפירוש הניתן בצלילים, כי אז — ורק אז — תהיה

They came to the classes from all over the country. Obviously most of them are from Jerusalem, but among the out-of-town guests—thirsty for knowledge—stand out teachers from Haifa, also from Galilee and the Emek. It should be recorded with satisfaction that many of those enrolled for the piano classes of Nadia Reisenberg visited the afternoon sessions of Jennie Tourel, and I am sure that no one was more glad than Miss Reisenberg herself. In her classes she emphasized the importance of a "singing" approach to piano playing…

Miss Nadia Reisenberg has a great gift of demonstrating the negative aspects of the pupil's playing, and in this way showing the undesirable, followed immediately by the desired, as performed by the teacher. The examples are done with such clarity that the pupil grasps immediately what requires correction. Teacher Reisenberg does not expect immediate results. Her main purpose is to stimulate the pupil—and through him everyone present, teacher and student alike—to think, to criticize, to study. In her opinion, one cannot separate analysis of technical material and the emotional-spiritual performance.

One word was not mentioned in all the classes I attended: "touch." It seems to me that teacher Reisenberg considers this concept to be the blend of all the positive characteristics of piano playing. If the pupil will use the pedal correctly (and she is very strict about this), if he will know how to color the playing with the proper dynamics, if he will "breathe" properly and understand the proper phrasing for each style…he may then achieve the Reisenberg touch which surprises everyone present so deeply.

Davar, August 24,1962

Tomorrow, my day looks like this: from 8:45 til 9:45, I give the first lesson; 9:45-10, on my way to the Municipality; 10-11, reception by request of (Mayor of Jerusalem) Teddy Kollek; 11-11:15, on the road back to school; 11:15-12:15, lesson #2; 12:30-2:45, lunch and reunion (rather short) with family; 3-6, more teaching; 6-6:30, first opportunity to go to the room where even the Tsar of Russia went on foot! And at 7, dinner, in order to remind me that I am a year older. Thank God I have to be reminded of it, for against all the rules of arithmetic (new methods or old), I just don't feel the piling up of a considerable number of years, which I am to keep whether I want to or not.

NR, letter home, 1962

NR and Yocheved Dostrovsky, Director, Rubin Academy, 1962

To our dear Nadia,
symbol of sincerety and
tendernes, the great artist
the unsurpassed teacher
with many thanks
for unforgettable hours
in Jerusalem

Yocheved Dostrovsky-
Kopernik
Jerusalem, Aug. 15, 1962

Nadia is a model of integrity and responsibility, a teacher of vast experience, and besides all this she has a radiant personality. Everyone is enthusiastic about her—and this is no small thing with musicians.

Yocheved Dostrovsky, 1962

THE Rubin Academy of Music in Jerusalem is again full of life although this is summer-time. Besides the course on dance, that goes on till August 12, and the course of dance therapy given by Miss Marian Chace of the St. Elizabeth Hospital, Washington, which ends today, there are the individual lessons in piano playing being given since the beginning of July to teachers and advanced students by Miss Nadia Reisenberg of the Mannes School of Music in New York. As of Sunday, Miss Reisenberg will teach open classes till August 20.

Miss Reisenberg describes the scheduled classes as a workshop. At each of the ten morning sessions students have agreed to serve as guinea pigs for her and she will comment on their playing for the benefit of the executant and an expected large audience of fellow students, teachers and musicians. Remembering previous workshops, we know that Miss Reisenberg's warm and stimulating personality will easily overcome the player's stage fright.

She has come here during her summer break, though giving private lessons to some 25 students a week is not exactly a holiday, but she knows every composition that is being studied because she plays it herself. A problem was the relatively limited repertoire students study here, and she would have liked to include more contemporary music.

The afternoon sessions will include varied fare. On August 9-13 composer Hanoch Jacoby will analyse works, on August 17-19 Bracha Eden and Alexander Tamir will disc[illegible]

Dearest children mine,

Nadia was at her brilliant best. You know her—she is a perfectionist. After the morning seminars she would practice herself, sometimes as much as three or four hours. To undertake at least twenty major works in a period of ten days is no joke, but only we knew how tired, how exhausted she was getting—her class could never guess it. She was alert, exciting, even for laymen like myself; her clarity of explanation and her superb performances, when showing how one should really make the piano sing, thrill everybody.

Newta, 1964

On Wednesday late afternoon, a farewell party was given to Nadia, who was variously described as "a miracle-worker," "a most remarkable musician," "a most profound and extraordinary teacher", "a fabulous pianist," and last but not least "a most understanding and sympathetic person," and "a warm and vibrant human being." All of this was said in Hebrew, Russian, and English without the need of simultaneous translation. The voices of the speakers (teachers and pupils alike) shook with excitement; eyes glistened with an occasional tear; Israeli and Russian hearts throbbed with excess emotion. Jokes aside, it was indeed a very touching and heartwarming occasion—even such hard-boiled Americans as Clara, Meir, and I were profoundly moved.

Newta, 1964

Jerusalem Post, 8/26/66

* * *

NADIA Reisenberg, the great pianist, is running a summer course in piano at the Rubin Academy; she also has been coming every two years since 1960 to serve the Academy. "This is supposed to be my summer vacation," she says, "It is rather more strenuous than a busman's holiday. I came to Israel incognito six years ago to see my sister: somehow I found myself working at the Academy. From then on, I have come every two years, which allows me a real holiday every second year."

Mrs. Reisenberg teaches three hours a day. Her master classes are attended by about 150 spectators. The performers are either young pianists or teachers; the audience, for the most part, consists of teachers, many of whom have come to Jerusalem from remote parts of the country. Two pianos are so arranged that the student at one of them, and Mrs. Reisenberg at the other, can watch each other's keyboards. After the student has played a piece, Mrs. Reisenberg plays it; the student plays it again, and tries to improve his or her rendition.

The audience participation is remarkable; the listeners obviously identify themselves closely with the student. When the student shows an improvement, the reaction is considerable. Altogether, in master classes and private lessons, Mrs. Riesenberg is devoting 25 hours a week to teaching at the Academy during the six-weeks summer course.

After the lesson, sitting in the shade of the giant tree of the Rubin Academy's courtyard, I ask what she thinks of the standard of our piano-playing.

"There is no such thing as an average standard," she answers. "They form a very heterogeneous melange. Some are young, some elderly, some have extraordinary talent but lack training, others are well-trained but are mediocre. Sometimes I find marvellous talent but it is not particularly well organised, other students may display remarkable proficiency but they lack artistic ability.

"I try to give them in six weeks what would take six months at home — I try to cram it into them. They are very, very keen — they work with almost overwhelming devotion and enthusiasm."

Mrs. Reisenberg ... celebrated pianist

of the states."

She concludes: "I would like to stay here for longer, so as to help the students — I think that something of what I teach does remain. I feel that I am really needed here. Israel means a great deal to me, especially Jerusalem, which must be one of the most beautiful cities in the world."

Mrs. Yocheved Dostrovsky, Director of the Rubin Academy, which is supported by the American-Israel Cultural Foundation, explains that the idea of the summer courses conducted by the two famous artists is to give teachers and young performers an op-

... work under great musical

... are arran...

Dearest Children,
I am sending you the result of an interview, which appeared in today's *Jerusalem Post.* The person who wrote it is obviously not a musician and the picture was taken two years ago!

NR, 1966

I went to Jerusalem to visit the Rubin Music Academy and the master classes of Nadia Reisenberg and Jennie Tourel. Nadia Reisenberg has a sense of humor and a sense of the dramatic. At the beginning she poses for a model lesson for the press-photographer, but then she sends him away, politely but firmly: publicity has its place, but the students' concentration must not be disturbed. The subject of the lesson is a sonata of Schubert, the problems are those of the spirit.

Two pianos are standing on the stage. At one sits the student (a teacher in daily life, perchance); at the second sits Nadia Reisenberg, attentive. The whole hall is listening carefully, the experienced teachers sit with the score in their hands. The piece comes to an end, and though it is not long, it sounded long. Nadia (everyone calls her that) begins her analysis:

"Schubert wrote 'Allegro' at the top, and this is usually played too quickly, but you played it too slowly. It sounded as though the music could not move forward..." She begins to illustrate, and immediately one sees that the monotony was not created by Schubert. "Usually I do not recommend extensive use of the pedal, but here you have no choice. What does the pedal give you? The pulse of life which your playing lacked."

Of course, a teacher has to be extremely tactful, particularly when correcting a pupil in front of an audience. "In this passage, you can introduce a touch of humor, to lessen the tension a bit. A little lighter, more mischievous. You sound as though you are doing a difficult job, with great responsibilities. Schubert is not like that, he is more..." and her fingers illustrate Schubert's airiness.

How does the pupil react to these perfect illustrations? Does he think, "I shall work until I too can play it like that," or does he say to himself, "I could never reach such perfection"? But even despair is a good lesson. There is no lack of first-class pianists in the world, so who is going to listen to second-class ones?

Nadia Reisenberg is an optimist, and this is a wonderful quality for a piano teacher. She loves to work, she loves to laugh, she loves to eat. At luncheon, we discussed her students of the morning, and I asked her where she found such patience. She laughed, and replied: If you don't have the patience of an angel you have no right to teach."

Michael Ohed, article in Hebrew language weekly, 1966

Left to right: Uri Mayer, Yuval Waldman, Isaac Stern, NR, Natan Brand, and Zita Finkelstein. The occasion was a benefit concert for the America-Israel Cultural Foundation, 1970

To my dear teacher, Nadia Reisenberg!

Please, remember me, and if you would like to recall who I am, just remember the boy whom you influenced not to move and twich his body during play.

Jerusalem Ramban st. 7 / August 1962 / Nathan Brand / We will always remember you in admiration and love

x x x x x

The teacher must be able to verbalize in addition to illustrating. You not only have to get at the root of a particular person's problem and point it out, you have to be able to explain in detail what is wrong, and why.

Some students have so much temperament that whenever they touch the piano it sizzles. Then I have to hold them back, to teach them gradations. If everything sizzles all the time, you don't even know it's sizzling.

Teaching is a calling — you have to want to show others what you know...

NR, 1981

April 11, 1974

Mrs. Nadia Reisenberg
165 West 66th Street
New York, N.Y. 10023

Dear Mrs. Reisenberg:

It turns out to be most appropriate that I address this communication to you during the Passover holiday. According to the Bible, it begins the New Year, and what better way for us at the America-Israel Cultural Foundation to express our highest admiration and deepest gratitude for your continuing contribution to the cultural life of our people, especially in Israel.

With this in mind, it is a privilege for me to inform you that our Board of Directors voted to present you with our first annual Tarbut (Culture) medal for the year 1974. This will take place at a special event at the Juilliard Theater in Lincoln Center on Wednesday, June 5th, 7:45 PM at the 18th (Chai) Anniversary Scholarship Celebration.

It will be my special joy to share with Mr. Isaac Stern, Co-Chairman of the Board of Directors, the opportunity of making the presentation to you on this occasion. We do hope that your schedule permits a positive response to this invitation.

With every good wish, I am

Sincerely yours,

Bernard Mandelbaum

BM:al

Isaac Stern, NR, Vera Stern, and Joseph Machlis

Israel Independence Day Celebration, 1968, with Andrei Sedych, Jenny Grey, and Clara

Have you ever held a child in your arms who falls asleep? He weighs a ton because he falls with dead weight. The moment you stiffen your arm, the same thing happens. You get a hard sound, and not even more volume because the sound will be much shorter, it won't carry. I use little motion, but I get a big sound because I use the weight of the arm—the free arm, not stiff. The wrist shapes the phrases, but the arm gives you the sonorities.

You can have the best ingredients in the world to make a cake, but just try to combine them in the wrong proportions, and see how wonderful your cake will be. You have to develop a sense of proportion and good taste in your students too.

A teacher must instill respect for the printed score because it's a language, like any other language. It has punctuation marks, it has long sentences and short ones, beginnings of phrases and ends of phrases...

NR, 1981

Last visit to Israel, 1981

Three Sisters

Newta, Clara, and NR

I don't know of another instance where three sisters were so close to each other, so tight, sharing so many common interests. They had very different temperaments, but they were alike in their attitudes—they were all extremely involved, enthusiastic, emotional. When one sister was worried about something, the other two were concerned as well, talking and discussing and reliving the story; whatever one sister was interested in, that was a compulsory interest for the two others. I seldom saw a difference of opinion: what one loved the others loved; they liked the same things (and people), and they disliked the same things. I've never seen that kind of unity in a family...

Andrei Sedych, 1985

I will sue
anyone swiping even
one picture!!!

AIR MAIL

Warning clipped to family photos by NR

We were often affectionately called, by American friends, the Chekov Sisters because of our Russian origins...

Clara, 1984

And sometimes we were called the Romanov Sisters because we loved to eat caviar—really good Russian caviar for Breakfast!

Newta, 1985

Clara and NR

NR and Newta

Newta and NR

NR and Clara

Clara and NR

Clara, NR and Newta

Clara, Newta, and NR

The three sister retained in their mental makeup much that was appealingly provincial. Musicians by the grace of God, and Bohemians by temporary choice, they were dependably bourgeois in their outlook and entirely reliable in their morals. They possessed a great deal of selfishness that was unconscious, but completely atoned for it by an abundance of charm and youthfulness which made them very attractive young women.

Meir Sherman, 1938

Clara and NR

NR, Newta, and Clara

Newta, Clara, and NR

Clara, Newta, and NR

NR and Clara

Clara, Newta, and NR

Clara, Newta, and NR

The sisters had an uncommon sense of attachment that is conceivable only among girls whose childhood passed without separation—in common experience of revolution, and hunger and pogrom...

Meir Sherman, 1938

New York Times, 7/22/79

MUSIC

Pianist to Celebrate 75th at Caramoor

By ROBERT SHERMAN

FAMILY MATTERS usually are not discussed in this column, but since my mother happens to be Nadia Reisenberg, the noted pianist, and since she further happens to be celebrating her 75th birthday month with an appearance at the Caramoor Festival in Katonah next Sunday, I decided that a few of her reflections on music might not be out of place here.

When the distinguished violist-conductor, Milton Katims, came to WQXR to pay tribute to her on her 70th birthday, he remarked that "in addition to her tremendous pianistic gifts, Nadia is primarily a musician of great depth, always examining what the composer was saying, always looking closely at what lies behind the printed page."

Perhaps because she makes a point of conveying that approach to her students, she is much in demand as a teacher (at the Juilliard School and the Mannes College, as well as privately). "I am convinced," she says, "that while technical standards are considerably higher among young pianists than they were 20 or even 10 years ago, musical values do often lag behind. That's why there are any number of marvelous players today, but not all that many astute musicians. That's also why, with my own students, I am more concerned with artistic development than technical excellence. In other words, I am prouder when a student can play a Brahms intermezzo with understanding and sensitive phrasing than I am if he can race through a Paganini-Liszt étude. Feeling for style and the achieving of a beautiful sound at the piano are more important elements to me than playing fast and loud, and they are infinitely more difficult to develop."

As a logical extension to preferring sensitivity over virtuosity, Nadia Reisenberg has long considered chamber music her greatest love. "It's the purist kind of music," she says, "and the player becomes more humble because he's less important than the music being played. In so many solo works we have to show off as performers — the acrobatic element is built into the pieces — but in the give and take of chamber music, you discover the most beautiful combinations of sounds. You sit there, and you're in heaven."

At Caramoor's Spanish courtyard, next Sunday at 5:30 P.M., Nadia Reisenberg will be appearing with Erick Friedman, the violinist, in a recital of sonatas by Mozart, Franck and Strauss. "We've had a wonderful association over the past several years," Mother recalled; "Erick is an especially flexible artist, we love trading ideas at rehearsal, and he takes precisely the same joy in chamber music that I do. I must say," she added, "I'm also very excited about playing in Caramoor again. It's such a gorgeous place, and being in an esthetic surrounding has always made a great difference to me. I don't know why, exactly, but I do better when I see lovely things around me!"

Erick Friedman and NR

Nadia was playing the Strauss Sonata with Erick Friedman. It had been a very unpleasant day, it had just stopped raining, the seats were wet, tempers were sort of wet. When they came to the slow movement (which she played particularly divinely), the sun suddenly broke through, and little birds not only began to sing, but started flying around the stage. It was a moment of pure magic...

Clara, 1984

Direct from her recital triumph upstairs, renowned pianist Alicia de Larrocha autographs some of her recordings that were then sold as premiums. Seated next to Miss de Larrocha at the WQXR broadcast table, left to right, are Philharmonic associate director of advertising and public information, Jack Murphy; pianist Nadia Reisenberg; and her son, WQXR program director, Robert Sherman.

When I was young, I used to think I would feel different when I got to be forty or fifty. Well, here I am at this ancient age of 75 and it still doesn't feel any different. It just doesn't. When you are involved with many things, you don't worry about whether you need this hour of rest in the afternoon, or that according to your age you should be slowing down. There just isn't time to think about it.

As I was coming down in the bus today, I thought to myself, "Well, here I am, it's almost time to summarize what my life has been like." And I asked myself a few questions: "What is it that you were really looking for in your life, how close did you get to achieving it, and if you had another life to live, what would you do differently?." And I gave myself the answers quite readily, because there they were: I wanted first of all to have a close family, a family I could love and a family that could love me. Second, whatever sort of talent I might have, I wanted to have the chance to develop it to its full potential (which would mean dedication, it would mean commitment, it would mean hard work), and as a person, I would like to have enough strength and enough integrity to fight for the things which I believe in. When I think of that, I almost accomplished it all. Of course we cannot accomplish *all*, but it's very close to being that which I set out to do, so I must consider myself the luckiest person in the world. If I had another life to live, I don't think I would want to have it any different. Well maybe I'd want to add a couple of hours to each day because I like to learn so much, and I have not enough time to do it all.

NR, 1979

The Mannes College of Music

MANNES CIRCLE SERIES

NADIA REISENBERG, Piano
DAVID GLAZER, Clarinet
CLAUS ADAM, Cello

Friday, March 2, 1979
Mannes Concert Hall
8:00 P.M.

157 East 74th Street, New York, New York 10021

75-ЛЕТИЕ Н. А. РЕЙЗЕНБЕРГ

Сегодня, 14 июля, знаменитой пианистке и выдающемуся педа-

Надежда Ароновна Рейзенберг

гогу Надежде Ароновне Рейзенберг исполняется 75 лет.

Цифра эта может поразить всех, кто по-прежнему называет Н. А. Рейзенберг "Надей". Не вяжется этот возраст с артисткой — молодой по внешности, полной энергии и жизнерадостности.

— Секрет этой, так называемой "молодости", — говорит Н. А., — кроется в работе. Чем старше я становлюсь годами, тем больше я работаю.

Работала она много всю свою жизнь. Надя Рейзенберг была первой, исполнившей по радио все 27 концертов Моцарта с оркестром WOR под управлением Валленштейна. Она много концертировала, была повторной солисткой Ньюйоркской филармонии, несколько сезонов выступала с Бостонским, Детройтским и многими другими симфоническими оркестрами в С. Штатах и в Европе. Существует несколько альбомов и большое количество долгоиграющих пластинок, записанных Н. А. Рейзенберг. К ее 70-летию фирма "Мюзикл Эридэдж" выпустила специальный альбом из лучших пластинок Нади Рейзенберг. Еще в этом году в "Нью Йорк Таймз" была помещена большая фотография пианистки, вместе с фотографией Артура Рубинштейна, и статья Гаролда Шонберга, который сравнивал исполнение одной и той же мазурки Шопена этими двумя артистами. Пластинки Нади Рейзенберг часто исполняются многими радиостанциями.

В жизни Нади Рейзенберг камерная музыка всегда занимала нье 29 июля в Карамур Фестивале состоится сонатный концерт Эрика Фридмана (скрипка) и Нади Рейзенберг (рояль). Они же дадут концерт камерной музыки 9 декабря в Карнеги Холл.

Свою педагогическую карьеру Н. А. Рейзенберг начала в Кэртис Институт в Филадельфии. Несколько лет она руководила семинарами для педагогов в Академии Рубина в Иерусалиме.

В настоящий момент Н. А. Рейзенберг преподает в трех ньюйоркских колледжах: в Квинс Колледж, в музыкальной школе Маннеса, в Джульярд Скул и заслуженно пользуется репутацией выдающегося педагога.

По случаю 75-летия пианистки завтра, в воскресенье 15 июля, радиостанция WQXR посвятит Наде Рейзенберг специальную программу из серии "Выдающиеся артисты". Интервьюировать Надю Рейзенберг будет ее сын, музыкальный директор станции Роберт Шерман. Интервью иллюстрирует музыка в исполнении Нади Рейзенберг.

Многочисленные друзья, поклонники и ученики шлют Н. А. Рейзенберг в день ее 75-летия самые искренние пожелания новых творческих успехов и еще долгих лет счастливой жизни.

АНДРЕЙ СЕДЫХ

PROGRAM

Sonata in B Flat for clarinet and piano	Johann Wanhal (1739-1813)
Allegro Moderato Adagio Cantabile Rondo Allegretto	
Sonata No. 2, Op. 120 (1894) for clarinet and piano	Johannes Brahms (1833-1897)
Allegro amabile Allegro appassionato Andante con moto-Allegro	
INTERMISSION	
Three pieces from Op. 83 for clarinet, cello, and piano	Max Bruch (1838-1920)
No. 4-Allegro agitato No. 5-Andante No. 7-Allegro vivace, ma non troppo	
Trio No. 4 in B Flat, Op. 11 for clarinet, cello, and piano	L. van Beethoven (1770-1827)
Allegro con brio Adagio Tema con Variazioni	

(75th Year)

(N.A. Reisenberg)

From an article in Novoye Russkoye Slovo by Andrei Sedych

Erick Friedman

New York Post, 10/12/77

By SPEIGHT JENKINS

The sweetness, power and solid intonation of Erick Friedman in his violin's high positions made his recital last night a remarkable one.

Held at the Theresa L. Kaufmann Concert Hall of the 92nd St. YM-YWHA, the recital united the talents of Friedman and pianist Nadia Reisenberg for an evening of Brahms, Beethoven and Richard Strauss. Though separated in years, the two made a wonderful team, and Miss Reisenberg showed that a great piano teacher can be a fine performing pianist, a situation not always the case.

All seemed a prelude to the vigorous and vital version of Beethoven's "Kreutzer" Sonata that concluded the evening. The outgoing, extroverted playing showed extensive rehearsal; rare is the violin-piano combination that can so beautifully effect such striking subito pianos (sudden soft moments).

Miss Reisenberg is not only a teacher, it should be noted, but the mother of radio station WQXR program director Robert Sherman, who took on a new role last night: a modest and unassuming page turner.

Carnegie Hall

1979-1980 SEASON

Sunday Evening, December 9, 1979 at 8:00

THE CARNEGIE HALL CORPORATION
presents

ERICK FRIEDMAN

Violin

NADIA REISENBERG

Piano

FRANCK — Sonata in A major for Violin and Piano
Allegretto ben moderato
Allegro
Recitative-Fantasia
Allegretto poco mosso

GRIEG — Sonata in C minor for Violin and Piano, Op. 45
Allegro molto ed appassionato
Allegretto espressivo alla romanza
Allegro animato

Intermission

STRAUSS — Sonata in E-flat major for Violin and Piano, Op. 18
Allegro ma non troppo
Improvisation: Andante cantabile
Finale: Andante—Allegro

For Mr. Friedman and Mme. Reisenberg
JOSEPH A. SCURO
International Artists' Management
111 West 57th St.
New York, N.Y. 10019

Steinway Piano

RCA, Monitor, Musical Heritage for Erick Friedman

Westminster, Monitor, Musical Heritage for Mme. Reisenberg

Mr. Friedman plays the Guarnerius del Geju: 1729, the Ex-Balokovic

This concert is made possible in part with public funds from the New York State Council on the Arts and the Department of Cultural Affairs of The City of New York.

The photographing or sound recording of any performance or the possession of any device for such photographing or sound recording inside this theater without the written permission of the management, is prohibited by law. Offenders may be ejected and liable for damages and other lawful remedies.

In recent years, I had the occasion to perform many sonata recitals with Nadia. I was so blessed with this opportunity to perform with such an artist—it was more meaningful to me than I can possibly say. It was an inspiration. Nadia Reisenberg was a complete musician, her musicianship transcended pianism, and certainly violinism; her musical knowledge of so many musical forms, of symphony, of opera, of so much of the literature, was breathtaking, it made each rehearsal an event—indeed I couldn't wait until the next one. Her experience, her professionalism was a fount of information. The only thing more impressive than her artistry was her warmth and tenderness as a human being. Finally, not only was Nadia the complete musician, she was the complete performer—she had that rare quality given to so few, whereby her inspired musicianship became enlightened performance. She loved music, and she loved people, and she loved to play for them—and they knew it. It was an honor for me to know her and to perform with her, and I cherish her memory. For me, she has not passed away, but gone forward into musical history.

Erick Friedman, 1983

New York Post, 12/10/79

Three scoops of violin romanticism is too rich

By SPEIGHT JENKINS

ONE OR EVEN two butterscotch sundaes are great at a sitting, but for even the most avid ice cream maniac, three would be a little much.

Which in musical terms is exactly what Erick Friedman and Nadia Reisenberg gave the Carnegie Hall audience last night at the former's Master of the Violin recital.

The three sonatas were the Franck, the C minor Grieg and the Strauss, all very sweet and very romantic pieces and all, oddly enough, composed in the two years 1886 and 1887. Their performances certainly revealed Friedman's mastery of the romantic violin literature, but he played so well that it would have been good to have heard him in more diverse music.

His tone was rich and full at all positions; his variety of vibrato, use of slides and long, singing line could scarcely be improved.

Intonation occasionally gave him trouble, but it seemed almost a function of his involvement in making music, by far the more important factor.

Certainly no one could have a better partner than Nadia Reisenberg. Unlike so many of her fellow piano teachers, Miss Reisenberg can really play. She displayed an infinite variety of tone and touch, a brilliant use of the pedal and the capacity to unite totally with the violinist while maintaining a strong, individual profile.

She has always been a great pianist; last night she was spectacular, with many memorable phrases even of music that was less than memorable.

Of the three pieces the Franck came off best because its quality romanticism was for once completely fulfilled. The Strauss though handicapped by a weak finale, has a fine Andante cantabile that was wonderfully shaped, while for all the expertise the Grieg never rose above an episodic joining of long-lined melodies.

Violins warm chilly Carnegie

By BILL ZAKARIASEN

Sunday was another important day for violinists in Carnegie Hall. For the matinee, veteran fiddler Ida Haendel, whose New York appearances have been regrettably rare of late, performed with the American Symphony Orchestra, while in the evening, Erick Friedman played in Carnegie for the first time in three years.

Friedman's recital of three romantic sonatas was not only beautifully geared to his talents, but also designed to show off the accompanist as virtuoso collaborator. He had a splendid partner in pianist Nadia Reisenberg, whose technical and intellectual mastery of her music gave ideal support. Reisenberg's well-known son, WQXR's Bob Sherman, served as efficient page-turner, but as he was wearing a bright red sport jacket, he was hardly unobtrusive.

New York Times, 12/11/79

Music: A Violin and Piano Partnership

By DONAL HENAHAN

THE number of violinists who play the instrument as sonorously as Erick Friedman belong to the Fingers of One Hand Club, meaning you do not have to count very high to run out of members. Mr. Friedman, who has been performing in public for 25 years, has long been outstanding for the sheer beauty of the sound he can draw from an ordinarily recalcitrant little four-stringed box. His recital at Carnegie Hall on Sunday night, with Nadia Reisenberg as pianist, demonstrated his continued mastery of the purely technical aspects of his craft.

How much of this attention to design was attributable to the Reisenberg presence at the keyboard would be difficult to say, but it was clear at any rate that the partnership worked. She took charge when the music required it, and never seemed less than an equal in the duo. Her underplaying of the sotto voce introduction to the Franck typified her thoughtful pianism all evening. Moreover, ignoring a couple of wrong notes early on, her part was handled with great aplomb in a program that called for an extraordinarily strong pianist throughout.

The flowers are sweet.
Leaves sway in the breeze
Glistening, the musical rays
Reflecting every shimmer!

Violins from trees,
Piano keys from ivory,
Music; nature; one.

Waters; reflecting
Music's dances on every crest,
Educating man.

Nature: What is it?
Moody and devastating,
Answered in music!

The flowers are sweet
Their fragrance is of music.
Life is wonderful.

Peter Sherman
at Carnegie Hall, 12/9/79

Sitting high above the city in an apartment conveniently located across from the Juilliard School, Nadia Reisenberg began by offering an overview of the *Barcarolle* as she feels it.

Luigi Pellettieri

Nadia Reisenberg

"If we start with the title, and if we take it literally, we see two elements: song and water. It is interesting that the work is so highly polyphonic. Chopin succeeded so marvelously in giving us the song throughout the piece while weaving around it a network of voices – sometimes melodic, sometimes as a response to a melodic statement. Along with this are all sorts of configurations which suggest water. In the A major section, beginning at bar 39, the inner voice is a sort of 'water-line' which must be kept constant. The monotony of those triplets, along with the bass line underlies the melody, which is not only in the upper voice at bar 42, but also in the inner voice at the reassertion of the theme.

"Chopin takes the same theme and treats it differently by giving it different surroundings. For example, the theme is first offered simply, in thirds with the bass line (another rolling motion suggestive of water).

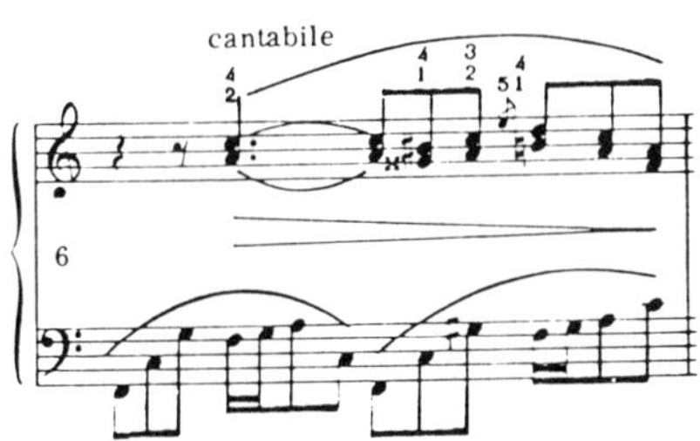

However, that same theme is offered again at bar 84 with the octaves in the bass and fuller chords in the right hand.

The theme at bar 62 is like a simple Italian street song, but watch what happens when Chopin puts the same theme into F♯ major at bar 93 with its rich harmonic structure and swells in the bass. These are like enormous waves.

There is so much more volume and richness to it. So it is not necessarily the tune itself which suggests or determines a feeling, but the setting and all that surrounds it. Of course the key of F♯ is so different from A major, so that offers a change in color, too.

Jumping ahead to bar 78 where Chopin writes the indication dolce sfogato, Reisenberg asked some Italian friends and gathered from them that the meaning has to do with breathing – perhaps suggesting a state of ecstasy or enchantment – a holding of breath. "That whole interlude is like a Nocturne for me It is so tender and lyrical that a person wants to hear it without any noise or distractions.

"I don't rush bar 93, in spite of the più mosso marking, because of the breadth of character in this section and because there must be a full, rich, open sound without ever being percussive."

Reisenberg feels the coda at bar 103 is noble and broader and should be played with generous rubatos, and a definite easing of the intensity of the preceding più mosso, as clearly indicated by the direction, tempo primo. She cautions about the text in bar 106 where the trills are in whole steps, and underscores what she feels to be an error in the Polish edition in bar 108 in the second triplet in the right hand: the last D♮ she believes ought to be a C♯ according to the natural harmonic progression of that passage.

Reisenberg does not see bar 110 as any kind of climactic moment, but merely a leggiero extension of what has come before, winding down to the calando. Her romantic and thoughtful reading of the work is obviously a combination of keen Russian instincts along with educated comparisons of the various editions. It is a pity that her beautiful on-the-spot performance cannot be shared with readers. Nadia Reisenberg has, however, recorded the *Barcarolle* for Musical Heritage. ■

Excerpted from an article by Carol Mont Parker in "Clavier" Magazine, April, 1983

NR with Dr. and Mrs. Peter Mennin

THE LIBRARY OF CONGRESS

THE GERTRUDE CLARKE WHITTALL FOUNDATION

THE JUILLIARD STRING QUARTET

ROBERT MANN and EARL CARLYSS, *Violins*
SAMUEL RHODES, *Viola* JOEL KROSNICK, *Violoncello*
and
NADIA REISENBERG, *Piano*

THE COOLIDGE AUDITORIUM
THURSDAY AND FRIDAY EVENINGS
APRIL 24 AND 25, 1980
AT 8:00 O'CLOCK

Washington Post, 4/26/80

Juilliard Quartet

Concerts by the Juilliard Quartet are usually distinguished, but the presence of pianist Nadia Reisenberg lent an uncommon luster to their program at the Library of Congress last night.

The years have in no way diminished her art. She plays with the same consummate musicianship that made her famous. Her fingers seem every bit as nimble, and her enormous experience shines through everything she does.

In the Mendelssohn D Minor Trio, it was her complete control of the idiom that was so delicious. Violinist Earl Carylss and cellist Joel Krosnick were with her every inch of the way in spirit but could not quite match the smoothness of her legatos in the first movement, or the lightness of her touch in the third. Reisenberg, on the other hand, had no trouble at all keeping up with their passion and power.

The Faure C Minor Piano Quartet brought an ideal balance of forces. Here, Reisenberg's marvelous touches of piano ornamentation were a foil for the broad sentiments of the strings, and the contrast and interplay of the instruments produced a splendid performance.

The concert began with a rollicking account of the Beethoven Opus 18 No. 6 String Quartet. It was the Juilliard at its best, which is saying a great deal.

—*Joan Reinthaler*

NADIA REISENBERG, Piano

ROBERT MANN, *Violin*
EARL CARLYSS, *Violin*
SAMUEL RHODES, *Viola*
JOEL KROSNICK, *Violoncello*

Tuesday Evening, April 29, 1980 at 8:00 p.m.

PROGRAM

Piano Trio No. 1 in D minor, Opus 49 — *Mendelssohn*

Molto allegro agitato
Andante con moto tranquillo
Scherzo: Leggiero e vivace
Finale: Allegro assai appassionato

Nadia Reisenberg, *Piano*
Earl Carlyss, *Violin*
Joel Krosnick, *Violoncello*

Sonata for Violoncello and Piano in G minor, Opus 19 — *Rachmaninoff*

Lento: Allegro moderato
Allegro scherzando
Andante
Allegro mosso

Nadia Reisenberg, *Piano*
Joel Krosnick, *Violoncello*

INTERMISSION

Piano Quartet in C minor, Opus 15 — *Fauré*

Allegro molto moderato
Scherzo: Allegro vivo
Adagio
Allegro molto

Nadia Reisenberg, *Piano*
Robert Mann, *Violin*
Samuel Rhodes, *Viola*
Joel Krosnick, *Violoncello*

* * *

The taking of photographs and the use of recording equipment are not allowed in the auditorium.

The Juilliard School welcomes your support to help continue this series of free concerts. Donations should be made payable to The Juilliard School, Lincoln Center, New York, New York 10023.

New York Times, 5/27/79

Recordings of Chopin's C-sharp minor Mazurka by Arthur Rubinstein and Nadia Reisenberg were paired in an informal survey made to see if performance differences between men and women could be identified.

'A FEW LAST WORDS': CUE CROSSWORD/By Maura B. Jacobson

Across

1 Having wings
6 Absent
10 —— Spee
14 Humpty's disaster
18 Take it easy
19 Fabricated
20 King of Midian
21 "Come —— faithful"
22 Pianist Reisenberg
23 Courtroom statement
24 Start of a quotation
26 Quotation continued
29 Pentacle
30 Fissionable item
31 Saarinen
32 Pouch
35 Coup for Goren
37 Paella
39 Is contrite
44 Post-marathon woe
46 Victorian oath
48 Do film-cutting
50 —— dull moment
51 Prevaricators
53 Film swashbuckler
55 Beaten track
57 Merino moms
58 "Did You Ever See —— Walking?"
60 Neighbor of Okla.
61 Raw recruits
63 Pumpernickels' kin
64 National Guard, e.g.
66 Woodland tribesman
68 Genetic
70 Rocks, at the bar
71 Quotation continued
74 Wonderment
77 Church councils
79 Watch winder
80 Garbed
82 Rue —— Paix
84 Arm span
86 Lobe site
88 Feels intuitively
89 "Beehive State"
90 Gene matter: abbr.
91 Moved sidewise
93 Chaucer pilgrim
94 Not more than
97 Egyptian sun disk
99 Mrs. Kovacs
101 Misfeasance
102 "Love's —— Lost"
104 Adjust the piano
106 Nick and Nora's pet
108 Trains on trestles, for short
109 Rug surface
111 Use swearwords
113 One of the Buddenbrooks
115 Quotation continued
123 End of the quotation
124 Bankruptcy
125 U.S. symbol
126 Greek peace goddess
127 Grafted, in heraldry
128 Aural
129 Tra followers
130 Gives silent assent
131 Darn it!
132 Direction for Greeley
133 Bloodhound's trail

Down

1 Florence's river
2 Table extender
3 Swit's costar
4 Tosser's choice
5 Puts on a pedestal
6 Current measurement
7 Hadrian's or Wailing
8 —— Rogers St. Johns
9 Dough additives
10 Made sleek
11 Bks. for research
12 Lessen
13 More phony
14 Berg
15 Astringent substance
16 One of the Redgraves
17 Was in the vanguard
21 Slanting
25 Ponti's spouse
27 Hay unit
28 Highway inn
32 Deli order
33 Vinegary
34 Author of the quotation
36 Swimmer Spitz
38 Sinewy
40 Mrs. Lloyd, once
41 Opus of the quotation
42 Logger's target
43 Impertinent retort
45 "Able was —— saw Elba"
47 Dragon constellation
49 Piedmont city
52 Lucifer
54 Torrent
56 Tiny island kingdom
59 Asia or Ursa
61 Made giggly sounds
62 Tizzies
65 Wrote a P.S.
67 Outside: prefix
69 Michaelmas daisy
72 Ethiopian lake
73 Mirror sight
75 Plantation pest
76 Detroit flops
78 Bumpkin
81 Arrow poison
82 Twofold
83 Vocalist James
85 Suffragist Carrie
87 Hester Prynne's stigma
91 Blasé state
92 In focus
95 Limber
96 Partner of true
98 Card games
100 Kind of collar
103 Hit man
105 Third-person custody
107 Farrah, Kate, and Jaclyn
110 Former queen of Italy
112 "On the Beach" author
114 Virtuoso Stern
115 "Gave proof —— the night. . ."
116 Hearken
117 Eternities
118 Word on a bulb
119 L.I. park
120 Give the eye to
121 —— Bator
122 Campus ordeal
123 Stannary product

Solutions to last week's puzzles appear on page 118.

The last public concert, Carnegie Recital Hall, October 3, 1981. Left to right: Vally Weigl, Kermit Moore, Barbara Peterson, Paul Doktor, and NR

Tender Care of a Cardiac Patient

6 am. A male attendant wakes you up by poking you in the ribs in order to take your temperature. You jump up in fright and he proceeds to stick a thermometer into your mouth.
7am. It takes you an hour to get over the shock, and at last you can close your eyes and try to sleep a little more, but you decide that a trip to the bathroom would be welcome, so you ring to get assistance since your doctor does not want you to try it alone. Not for another few days. You wait, and wait. At last the attendant comes, and with a sneer says "Must you have someone help you to make the few steps to the bathroom?" You either would like to curse him good and strong, or you keep your mouth shut. I choose the latter. The function accomplished, you ring again to have assistance to get back to bed. At last the desired rest!!!

But by that time it is 8 am and breakfast is brought in. Right in the middle of your Rice Krispies (the bread is like rubber and uneatable) a female nurse comes in to draw blood. Another beauty mark to add to all the other black, blue, brown and even, believe it or not, green colors on your arms. When you are lucky, it is a short procedure, but God save you from some who make you jump with pain when they search for a "good" vein. Now, now I can finish breakfast —a piece of farmer cheese. One bite and I spit it out—salty like hell. Because of all the excitement, a need for another trip to the bathroom arises (see previous experience).
9 am. At last I'm in bed. Ann, the Sweet Angel (I will write about her separately) brought me the N.Y. Times, and her concern and tenderness touch me deeply. Now I can rest at last. Rest I did—thirty minutes in all. Still to come: checking of blood pressure (in the middle of lunch) and a cardiogram.

What is this? Hard labor? Punishment for sins? No, this is the tender care I stated in the beginning. But after all it could be worse. I am still alive and kicking!!

NR, handwritten in Beth Israel hospital, 1982

Last photos of NR, April, 1983

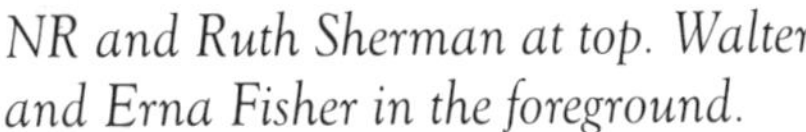

NR and Ruth Sherman at top. Walter and Erna Fisher in the foreground.

NR with David Sherman

The real values in life have little to do with physical disability or discomfort. When I ask myself what are the things that matter, my thoughts immediately turn to those I love—my family and my friends—the principles I believe in and try to live up to. To enjoy life I need two things—a brain to reason with and a heart to feel with. To love and to be loved; to enjoy the sun, the air; to try to be of some use in making the world a better place. To read books is already so very much!

NR, letter to Erna Fisher, 1980

What a joy and deep satisfaction to make music with Nadia. It was such a great feeling of partnership, knowing that at all times we somehow knew what the other was about to express and did it together—hardly any need to request a particular shaping of a detail; with Nadia, things had a way of falling into place just the right way! Her touch, her musical line were inimitable—what a wonderful artist, what a formidable musician she was!

Paul Doktor, 1985

I take great comfort and joy in knowing that her last birthday in this world was celebrated in my house, with the three inseparable sisters, Nadia, Newta, Clara. Wagner may have had leitmotifs for different characters in his operas, and I'm no Wagner, but many years ago Nadia and I, whose repertoire together and separately is completely classical, fell in love with a French song which was popular at that time, "La Vie en Rose." I personally adopted that tune as her leitmotif. It suited her perfectly, La Vie en Rose, because she enjoyed life, she used life, she brought love to other people and she was loved *by* other people. At her last birthday, everything was roses, the tablecloth had roses, even the napkin had roses on it, she was full of spirit and joy. And on that birthday, when we played some pieces together—we didn't do anything serious, just light things like "Summertime" and "Midnight Bells", all with lots of feeling—and, of course, we played "La Vie en Rose."

Clara, 1984

I saw her the day before she passed away. We had lunch together, and she was telling me that she's still a very sick person. "I have to be careful, I don't feel too strong." But she was very happy, in an excellent mood, and even though she talked about sickness, she was cheerful and optimistic. We spent a quiet afternoon talking...I miss her very much. We all do...

Andrei Sedych, 1985

11 June 83

Dear Bob,

The beauty of your mother's spirit lives on in our hearts, and her music is imperishable in our memory.

My deepest sympathy and profound condolences.

In friendship, always,

Gene

(List)

Nadia and I were colleagues and friends. At Mannes, as I know it was at every place that she taught, she was regarded as one of the great columns that held up the institution. She was always available for advice if you needed it—most of the time I got a lot more than I asked for—and in all of these discussions, she was invariably focused on artistic integrity and on a personal concern for the people with whom she was involved. And what is more, she insisted I must do the same. During her later years, her intense devotion to sharing her insights and her skills with the next generation deprived the rest of us of many significant musical experiences, because she was an astounding pianist. Everyone here must have personal memories of particular performances of Nadia's that meant a great deal to us. These last months, Nadia and I spoke by phone rather often; my initial questions about the general state of her health were always brushed aside with some kind of general statement that she was coming along, or feeling better, and she always stressed that she was so anxious to get back to a full teaching schedule. That last was a Reisenberg theme which was subjected to innumerable variations. I launched the idea that, because of her illness, a leave of absence of some sort might be helpful...that idea never reached deep water; she sank it at the dock. She let me know with great clarity and great sureness that these were *her* students, that she had brought them along this far, that their artistic development was *her* responsibility, and that she would work it out. And when the smoke had cleared, she *did* work it out, and certainly the lives of the students were much the richer for the extra time they had with Nadia.

Whenever pianists and pianistics of this century are discussed, Nadia's accomplishments must be pivotal. Hers was one of our century's richest and most giving lives in music. A source of great richness has gone out of our lives, but the value of our having known Nadia will surely remain with us, and will continue to furnish us with a wonderful model of a musical life that was generous and loving.

Charles Kaufman, President, Mannes, College of Music, 1983

July 14th 1980

To my entire family — sisters, children grandchildren and a few selected very close friends, who should not find it difficult to recognize themselves.

To-day I reached my 76th Birthday and my heart is overflowing with gratitude for every day of my life. As I have but a very few years left (if at all) I try to look back and evaluate what my life was like, and what I have achieved, ~~and~~ I feel that I have little to be ashamed of and much that I can be proud of. I value above anything else — love and work, both of which sustained me and made every day a valuable experience for which I am extremely grateful. I hope that I will be granted the biggest present of all — clarity of mind till the very end and ability to feel, to retain my enthusiasm and love and gratitude for a very happy life.

I love you all and wish every one of you to have a rich and rewarding life such as mine has been

Mama, Mum, Didi

Nadia often said that she wasn't ready yet to play the harp in heaven. I have a visual picture of her sitting in heaven right now—sitting, not floating like an angel, because her favorite position was sitting at the piano. I don't see any harps, but I see Nadia sitting, and watching over us, and giving orders...

Clara, 1984